Galina Filipuk, Andrzej Kozłowski
Analysis with Mathematica®

Also of Interest

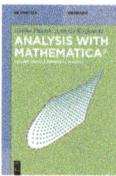

Analysis with Mathematica®. Volume 1: Single Variable Calculus
Galina Filipuk, Andrzej Kozłowski, 2019
ISBN 978-3-11-059013-5, e-ISBN (PDF) 978-3-11-059014-2,
e-ISBN (EPUB) 978-3-11-059015-9

Computational Physics. With Worked Out Examples in FORTRAN and MATLAB
Michael Bestehorn, 2018
ISBN 978-3-11-051513-8, e-ISBN (PDF) 978-3-11-051514-5,
e-ISBN (EPUB) 978-3-11-051521-3

Mathematica und Wolfram Language.
Einführung – Funktionsumfang – Praxisbeispiele
Christian H. Weiß, 2017
ISBN 978-3-11-042521-5, e-ISBN (PDF) 978-3-11-042522-2,
e-ISBN (EPUB) 978-3-11-042399-0

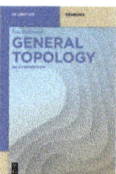

General Topology. An Introduction
Tom Richmond, 2020
ISBN 978-3-11-068656-2, e-ISBN (PDF) 978-3-11-068657-9,
e-ISBN (EPUB) 978-3-11-068672-2

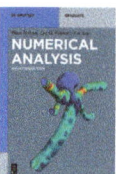

Numerical Analysis. An Introduction
Timo Heister, Leo G. Rebholz, Fei Xue, 2019
ISBN 978-3-11-057330-5, e-ISBN (PDF) 978-3-11-057332-9,
e-ISBN (EPUB) 978-3-11-057333-6

Galina Filipuk, Andrzej Kozłowski

Analysis with Mathematica®

Volume 2: Multi-variable Calculus

DE GRUYTER

Mathematics Subject Classification 2010
97I20, 97I60, 53A04, 53A05, 97N80

Authors

Dr. hab. Galina Filipuk
University of Warsaw
Faculty of Mathematics, Informatics and
Mechanics
Banacha 2
02-097 Warsaw
Poland
filipuk@mimuw.edu.pl

Dr. Andrzej Kozłowski
University of Warsaw
Faculty of Mathematics, Informatics and
Mechanics
Banacha 2
02-097 Warsaw
Poland
akoz@mimuw.edu.pl

ISBN 978-3-11-066038-8
e-ISBN (PDF) 978-3-11-066039-5
e-ISBN (EPUB) 978-3-11-066041-8

Library of Congress Control Number: 2020944402

Bibliographic information published by the Deutsche Nationalbibliothek
The Deutsche Nationalbibliothek lists this publication in the Deutsche Nationalbibliografie;
detailed bibliographic data are available on the Internet at http://dnb.dnb.de.

Cover image: Created by the authors with the help of Mathematica®
Typesetting: VTeX UAB, Lithuania
Printing and binding: CPI books GmbH, Leck

www.degruyter.com

Contents

Preface

This volume is devoted to analysis of multi-variable functions. Functions of one variable were considered in Volume 1 [5]. We continue to demonstrate that Mathematica®[1] can be used effectively as an aid in solving mathematical problems, or at least in discovering their solutions. As in the first volume, we leave most proofs and many other details to several recommended textbooks, and we also assume that the reader knows the basics of the Wolfram Programming Language™. Also as in the first volume, we use some problems from the lecture notes [12] for the course of analysis for computer science students given at the University of Warsaw during 2011–2018. We refer to the preface of Volume 1 for more information.

When writing Volume 1, we used Mathematica®'s version 11, whereas this volume is based on version 12 which was released while writing this book. Although there are some differences between the two versions, almost everything described in Volume 1 works in the same way in version 12.

Although there exists several books on calculus with Mathematica®, we hope that our two volumes offer a somewhat different perspective and will inspire the reader to experiment with other topics in analysis, which we had to leave out. This book is primarily intended for educational purposes, but we believe that some parts of it can be of interest to researchers.

---

1 http://www.wolfram.com/

https://doi.org/10.1515/9783110660395-201

1 Continuity of functions and mappings

In this book we shall deal with mappings from a subset $U \subset \mathbb{R}^n$ to $\mathbb{R}^m$ for arbitrary positive integers m and n, where at least one of them is greater than one. Such mappings f are given as collections of m functions $(f_1, f_2, \ldots, f_m)$, where $f_i : U \to \mathbb{R}$ is a function of n variables. Usually, for $m = 1$, they are called *functions* and for $m > 1$ *mappings* (*maps, vector-valued functions*).

In this chapter we first deal with $\mathbb{R}^n$ and introduce the notions of metric, norm, inner product, and discuss relations between them. Then we define the notions of limits, topology, and continuous mappings. We show how to visualize mappings in Mathematica® and study their continuity. As an application we illustrate how continuous mappings can be used to analyze the topological properties of sets.

1.1 Metrics

Recall that a *metric space* is a set X together with a map $d : X \times X \to \mathbb{R}_{\geq 0}$ with the following properties [15, p. 650]:
1. $d(x, y) = 0$ if and only if $x = y$;
2. $d(x, y) = d(y, x)$;
3. $d(x, z) \leq d(x, y) + d(y, z)$ (triangle inequality).

Mathematica® has many built-in functions containing the word "distance" as one can check by typing

> *In[·]:=* ?*Distance

For instance, EuclideanDistance, HammingDistance, GraphDistance, Travel-Distance, and many others. However, not all of them refer to a distance in our sense. The most general, applicable to any set X, is the function BinaryDistance:

> *In[·]:=* BinaryDistance[x, y]
> *Out[·]:=* Boole[x ≠ y]

In this metric the distance between any element and itself is 0, and the distance between any unequal elements is 1. It is easy to verify the triangle inequality by hand. If we try to use Mathematica® for this purpose, we obtain a somewhat strange result:

> *In[·]:=* FullSimplify[BinaryDistance[x, y] +
>   BinaryDistance[y, z] >= BinaryDistance[x, z]]
> *Out[·]:=* x ≠ y || x == z || y ≠ z

Of course, this relationship should evaluate to True. However, Mathematica® cannot simplify such expressions unless it knows that the symbols represent real (or complex) numbers:

https://doi.org/10.1515/9783110660395-001

In[·]:= FullSimplify[%, Assumptions -> Element[_, Reals]]
Out[·]:= True

The most important distance for us will be the EuclideanDistance. It is normally defined on the Cartesian product $\mathbb{R}^n$, but since **Mathematica**® by default works over the complex numbers, it computes the distance in $\mathbb{C}^n$:

In[·]:= ed = EuclideanDistance[{x, y}, {a, b}]
Out[·]:= $\sqrt{\text{Abs}[-a + x]^2 + \text{Abs}[-b + y]^2}$

One way to deal with this is to apply the function ComplexExpand, which by default assumes that all parameters are real:

In[·]:= ComplexExpand[ed]
Out[·]:= $\sqrt{(-a + x)^2 + (-b + y)^2}$

We can also use ComplexExpand to see that the distance on the set $\mathbb{C}^n$ is just the Euclidean distance on $\mathbb{R}^{2n}$:

In[·]:= FullSimplify[ComplexExpand[Sqrt[Abs[-a + x]^2 +
            Abs[-b + y]^2], {a, b, x, y}]]
Out[·]:= $\sqrt{\text{Im}[a - x]^2 + \text{Im}[b - y]^2 + \text{Re}[a - x]^2 + \text{Re}[b - y]^2}$

Note that ComplexExpand treats all variables as real unless they are listed in the second argument.

Other useful distances on $\mathbb{R}^n$ are ManhattanDistance and ChessboardDistance:

In[·]:= ManhattanDistance[{x, y, z}, {a, b, c}]
Out[·]:= Abs[-a + x] + Abs[-b + y] + Abs[-c + z]

In[·]:= ChessboardDistance[{x, y, z}, {a, b, c}]
Out[·]:= Max[Abs[-a + x], Abs[-b + y], Abs[-c + z]]

As can be seen from the above, the ManhattanDistance between two points is the sum of the distances between their projections on the coordinate axes, while the ChessboardDistance is the maximum of these distances. It is easy to show that both satisfy the triangle inequality.

A very important consequence of having a metric on a set X is that it defines a topology on X. In order to define a topology on X, we need the notion of an *open set*, and to define open sets, we need the notion of an open ball. The *open ball* with center at x_0 of radius $r > 0$ is the set $B(x_0, r) = \{x \in X \mid d(x, x_0) < r\}$. A subset $Y \subset X$ is called a *neighborhood* of $y \in Y$ if there is an $r > 0$ such that $B(y, r) \subset Y$. A subset $U \subset X$ is said to be open if it is a neighborhood of its every point. *Closed sets* are the complements of open sets. Clearly, the empty set and the whole space X are both open and closed sets.

To understand open sets with respect to a metric, we need to understand the open balls. The following function draws a ball of a given radius for a given metric:

In[·]:= `ball[metric_, r_] := RegionPlot[metric[{x, y}, {0, 0}]`
`< r, {x, -2*r, 2*r}, {y, -2*r, 2*r}]`

Note that since `RegionPlot` never draws objects of dimension less than two, it will not correctly represent `ball[BinaryDistance, 1]`, which is a point (actually the picture is empty). Let us consider the balls in the following metrics:

In[·]:= `{ball[EuclideanDistance, 1], ball[ManhattanDistance, 1],`
`ball[ChessboardDistance, 1]}`

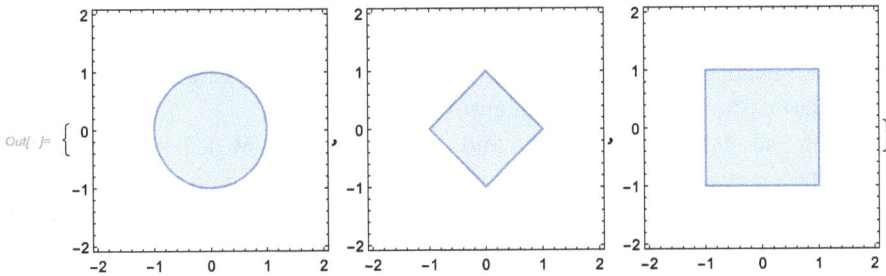

Figure 1.1

We can now see that although the open balls in these metrics have different shapes, a ball in each metric contains and is contained in a ball of any other metric:

In[·]:= `Show[(ball[#1[[1]], #1[[2]]] & ) /@`
`{{EuclideanDistance, 1}, {ManhattanDistance, 1/2},`
`{ChessboardDistance, 1/4}, {EuclideanDistance, 1/8}}]`

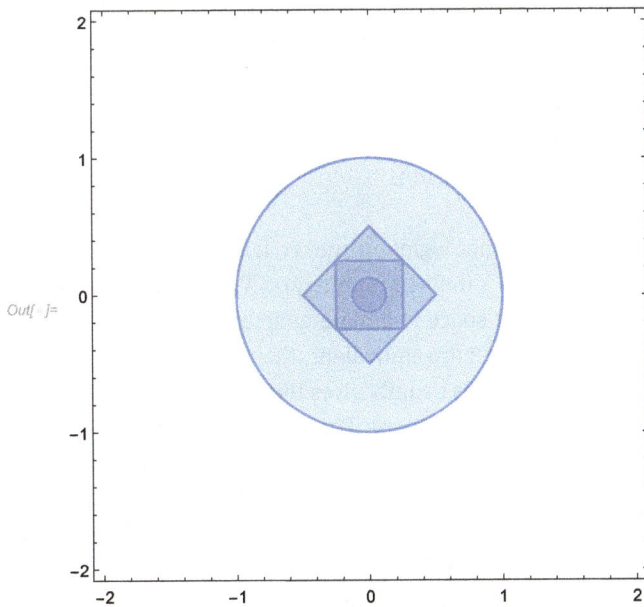

Figure 1.2

This implies that any open set with respect to any of these metrics is also open with respect to any other of these three metrics.

We say that two metrics d_1 and d_2 on X are *equivalent* if there exist constants $c_1 > 0$ and $c_2 > 0$ such that $c_1 d_1(x,y) \le d_2(x,y) \le c_2 d_1(x,y)$ for all $x, y \in X$. It is easy to see that all the metrics above are equivalent. Clearly, the binary metric is not equivalent to them.

In a metric space (X, d) one can define the notion of the limit of a sequence completely analogously to the way it was done in the case $X = \mathbb{R}$. In fact, we can reduce this definition to the definition of the limit of a sequence of real numbers: x is the limit of a sequence x_n if $\lim_{n \to \infty} d(x, x_n) = 0$. Limits of sequences are unique and have properties analogous to those of sequences in $\mathbb{R}$.

Given two metric spaces (X_1, d_1) and (X_2, d_2), we can now define the notion of *continuity of a map* $f : X_1 \to X_2$. Namely, we say that f is continuous at a point x if for any sequence x_n in X_1 whose limit is x, we have $\lim_{n \to \infty} f(x_n) = f(x)$, where the limit is taken in X_2 with respect to d_2. In particular, taking $X_2 = \mathbb{R}$ and d_2 the Euclidean metric, we obtain the notion of continuity of a real-valued function defined on any metric space. A mapping defined on a metric space X is continuous if it is continuous at each point of X.

The notion of continuity can be described also without referring to limits of sequences. Namely, let $f : X \to Y$ be a mapping between two metric spaces. Then f is continuous if and only if for every open (closed) set $A \subset Y$ the inverse image $f^{-1}(A)$ is open (closed). This definition does not use the notion of metric at all, and it can be applied to more general objects than metric spaces. More precisely, by a *topological space* we mean a set X together with a family Ω of subsets (called open sets) with the following properties:

1. The empty set and the set X belong to Ω;
2. Arbitrary sums of elements of Ω are in Ω;
3. Finite intersections of elements of Ω are in Ω.

In general, there can be many different topologies on one set. It is easy to check that the open sets defined earlier by a metric on X define a topology on X, but not all topologies arise from metrics. Thus a topological space is a more general object than a metric space. When two metrics on the same set are equivalent, their underlying topologies are the same (open sets are the same). This usually gives the easiest way to show that two metrics on the same space are not equivalent. For example, the discrete metric (BinaryDistance) on $\mathbb{R}^n$ is not equivalent to the EuclideanMetric on $\mathbb{R}^n$. Clearly, in the topology induced by the discrete metric, every one-point set is open (it is the open ball with center at that point and radius 1/2) while in $\mathbb{R}^n$ with the Euclidean metric one point sets are never open. The topology induced by the discrete metric is known as the *discrete topology* and is the only topology in which all sets are simultaneously open and closed.

1.1.1 Example: the railway metric

Often by analyzing the geometry of the unit open balls, one can determine that two different metrics are not equivalent. For example, consider the following metric d on $\mathbb{R}^2$, which we call the "railway" metric and which can obviously be generalized to $\mathbb{R}^n$. To define this metric, we choose a point w in $\mathbb{R}^2$ which will serve as the "node". In the illustration below, w is chosen to be the point $(0,0)$. We assume that all railway lines are straight lines that pass through the node. Therefore, if two points lie on such a line, the distance between them is just the usual Euclidean distance. If they do not, then the only way to travel from one point to the other is through the node and therefore the distance between them is the sum of the distances between each of them and the node. Let us consider the unit open ball with center at the point $p = (p_1, p_2)$ in this metric. There arise three distinct cases depending on the distance between w and p. If $w = p$, it is obvious that the unit ball in our metric is just the usual unit ball in the Euclidean metric. If the distance between w and p is less than 1, then the unit ball in the railway metric is the union of the Euclidean ball and an interval. This ball is centered in w and has radius equal to $1 - d(p,w)$. The interval is centered at p and it has length 2 (some points may lie inside the ball). If the distance between w and p is greater than 1, only the interval will remain. Clearly, this set is one-dimensional, and it cannot contain any two-dimensional Euclidean open ball, hence it is not open in the Euclidean metric. Therefore, the metrics are not equivalent.

To visualize a unit open ball in the railway metric in Mathematica®, we define the distance $d(x,y)$ equal to EuclideanDistance[{x,y},p] if (x,y), p, and w lie on the same line, otherwise it is EuclideanDistance[{x,y},w]+EuclideanDistance[w,p] (since one can travel from one point to another only by going along the line passing through the node). However, RegionPlot used in the straightforward way does not give us a correct picture. The reason is mentioned before: the function RegionPlot plots only generic points and omits objects of lower dimensions. So, in order to obtain a full picture, we need to add manually the missing part.

```
In[·]:= Manipulate[RegionPlot[If[y*p[[1]] == p[[2]]*x,
        EuclideanDistance[{x, y}, p], EuclideanDistance[
        {0, 0}, p] + EuclideanDistance[{x, y}, {0, 0}]] <= 1,
        {x, -2, 2}, {y, -2, 2}, Epilog -> {Point[p], Text[
        "w", {0.05, 0.05}], Text["p", p + {0.1, 0.1}], Point[
        {0, 0}], If[p != {0, 0}, {Directive[Opacity[0.5],
        Red], Line[{p - p/Norm[p], p + p/Norm[p]}]}, {}],
        Dashing[{0.05, 0.05}], Circle[{0, 0}, 1]},
        AspectRatio -> Automatic, PlotStyle -> {Opacity[0.5],
        Red}], {{p, {0, 0}}, Locator}]
```

Out[ ]=

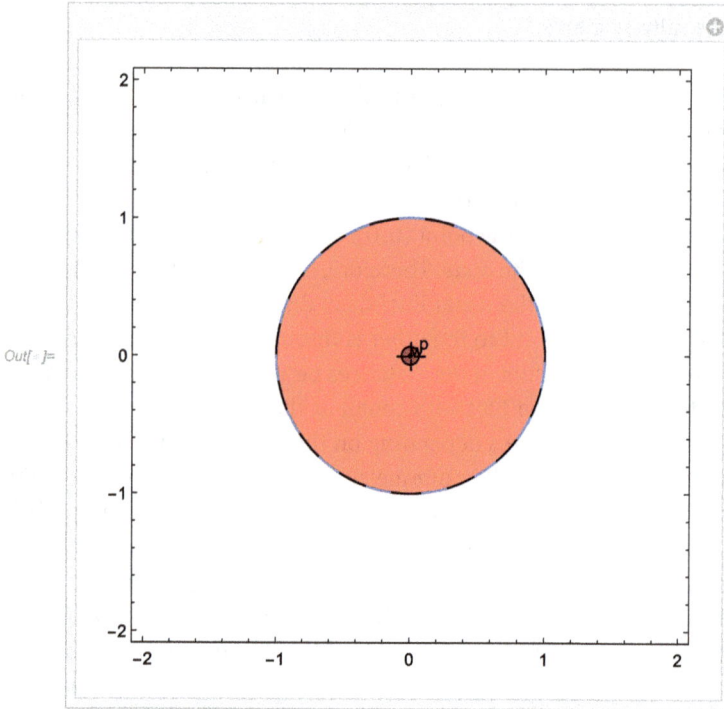

Out[ ]=

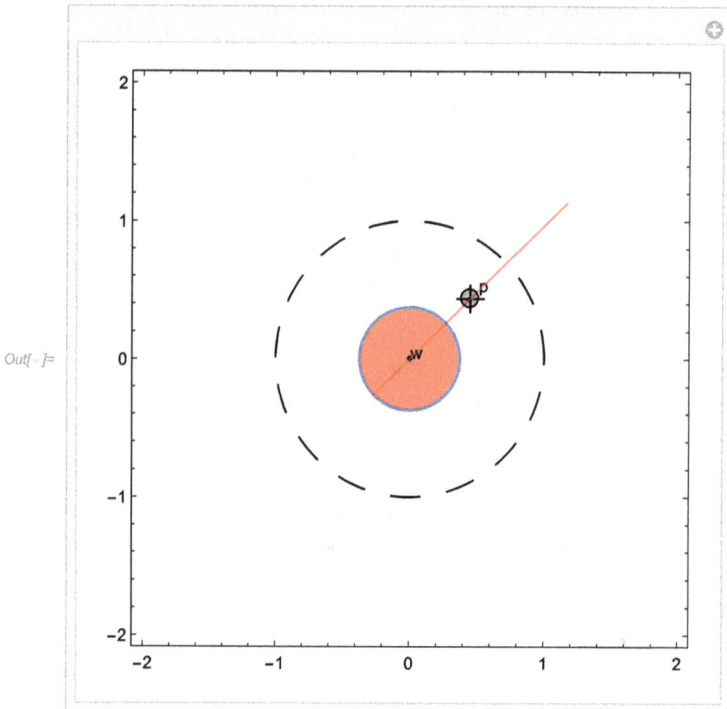

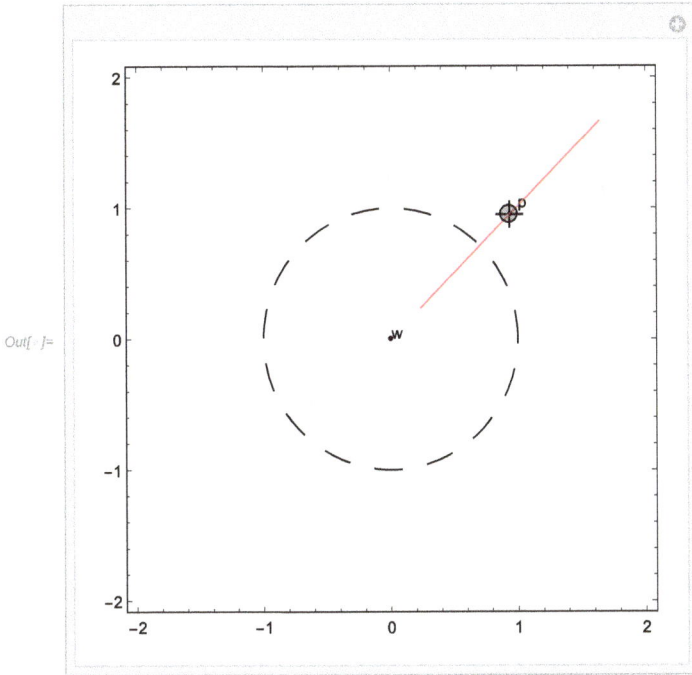

Out[ ]=

Figure 1.3

1.2 The vector space $\mathbb{R}^n$

Recall that the set $\mathbb{R}^n$, the n-fold Cartesian product of $\mathbb{R}$, that is, the set consisting of n-tuples $x = (x_1, x_2, \ldots, x_n)$, has the structure of a vector space, that is, its elements can be added and multiplied by real numbers (scalars), so that the usual rules are satisfied (see [15, p. 645]).

A vector in Mathematica® is a list of numbers, e. g.,

In[ ]:= {0, 1, 0.5}
Out[ ]:= {0, 1, 0.5}

One can check whether something is a vector or not with the help of the function VectorQ:

In[ ]:= VectorQ[{1, 2, Pi}]
Out[ ]:= True

In[ ]:= VectorQ[{1, 2, {Pi}}]
Out[ ]:= False

Vectors of the same dimension can be added and multiplied by scalars:

In[ ]:= {2, 3, 4} + {1, 2, 3}
Out[ ]:= {3, 5, 7}

In[·]:= a*{2, 3, 4}
Out[·]:= {2 a, 3 a, 4 a}

Vectors in 2 and 3 dimensions are graphically represented as arrows. There are two equivalent approaches. One is to view all vectors as arrows attached at the origin. Such a vector is called the position vector of the point where the head of the arrow ends. The other regards a vector as a class of arrows with the same length.

In the following interactive illustration, we start with two given vectors *v* and *w* and two scalars *a* and *b*. On the picture we can see vectors *av*, *bw*, and their sum.

In[·]:= Manipulate[Show[Graphics[{Arrow[{{0, 0}, a*v}],
        Text[a*"v", a*v + {0.1, 0.1}], Arrow[{{0, 0}, b*w}],
        Text[b*"w", b*w + {0.1, 0.1}], Arrow[{{0, 0},
        a*v + b*w}], Text[a*"v" + b*"w", a*v + b*w +
        {0.1, 0.1}]}], PlotRange -> {{-1, 1}, {-1, 1}}],
        {{v, {1/2, 1/2}}, {-1, -1}, {1, 1}, Locator},
        {{w, {-2^(-1), 1/2}}, {-1, -1}, {1, 1}, Locator},
        {{a, 1, a}, -2, 2}, {{b, 1, b}, -2, 2}]

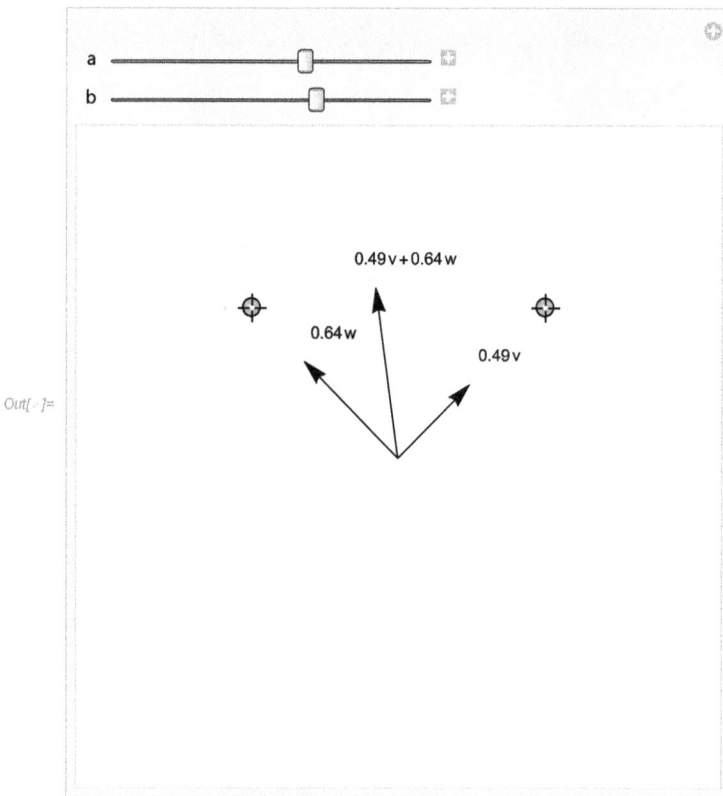

Figure 1.4

1.3 Norms

In the classical case of vectors in 2 or 3 dimensions, we have the notion of the length or magnitude of a vector, which is just the Euclidean distance between the end point of the position vector determined by the vector and the origin of coordinates. A generalization of this concept is called a norm. Recall from [5, Chapter 6] that a vector space V is called a *normed linear space* if there is a function (a norm) $\|\cdot\| : V \to \mathbb{R}_{\geq 0}$ with the properties

1. $\|x\| = 0$ if and only if $x = 0$;
2. $\|\lambda x\| = |\lambda| \|x\|$ for $\lambda \in \mathbb{R}$;
3. $\|x + y\| \leq \|x\| + \|y\|$.

On a normed linear space, one can always define a metric by the formula

$$d(x, y) = \|x - y\|.$$

However, not every metric can be obtained from a norm in this way. For example, the discrete metric cannot be obtained in this way since for any nonzero x we have $d(2x, 0) = 1$ but this is not equal to $2d(x, 0) = 2$ (see the second property of the norm).

The *Euclidean norm* is given by the formula

$$\|(x_1, \ldots, x_n)\| = \sqrt{\sum_{i=1}^{n} x_i^2}.$$

In Mathematica® the Euclidean norm is denoted by Norm. For example:

In[·]:= Norm[{1, 1, 1}]
Out[·]:= $\sqrt{3}$

As we already know, in symbolic expressions Mathematica® assumes that all the variables are complex:

In[·]:= Norm[{a, b, c}]
Out[·]:= $\sqrt{\text{Abs}[a]^2 + \text{Abs}[b]^2 + \text{Abs}[c]^2}$

If we deal with real numbers, we can use ComplexExpand:

In[·]:= ComplexExpand[%]
Out[·]:= $\sqrt{a^2 + b^2 + c^2}$

Another way is, for example,

In[·]:= Sqrt[Abs[a]^2 + Abs[b]^2] /. Abs[x_] -> x
Out[·]:= $\sqrt{a^2 + b^2}$

The standard Euclidean norm is actually a special case of a *p*-norm, defined for any $p \geq 1$ as follows:

In[·]:= ComplexExpand[Norm[{x, y, z}, p]]
Out[·]:= $\left((x^2)^{p/2} + (y^2)^{p/2} + (z^2)^{p/2} \right)^{\frac{1}{p}}$

For $p = 1$, we get the Manhattan norm

> *In[·]:=* Norm[{x, y, z}, 1]
> *Out[·]:=* Abs[x] + Abs[y] + Abs[z]

The p-norm is also defined for $p = \infty$, when it gives the chessboard norm:

> *In[·]:=* Norm[{x, y, z}, Infinity]
> *Out[·]:=* Max[Abs[x], Abs[y], Abs[z]]

For the simplest cases, **Mathematica**® can verify the triangle inequality: for u, v, p, q in $\mathbb{R}$, one gets

> *In[·]:=* Reduce[ForAll[{u, v, p, q}, Element[{u, v, p, q},
>   Reals], Norm[{u, v} + {p, q}] <= Norm[{u, v}]
>   + Norm[{p, q}]]]
> *Out[·]:=* True

> *In[·]:=* Reduce[ForAll[{u, v, p, q}, Element[{u, v, p, q},
>   Reals], Norm[{u, v} + {p, q}, 1] <=
>   Norm[{u, v}, 1] + Norm[{p, q}, 1]]]
> *Out[·]:=* True

The reader can check that **Mathematica**® also verifies the triangle inequality for $p = \infty$.

Equivalence of norms is defined analogously to that of metrics. However, unlike in the case of metrics, we have the following useful theorem.

Theorem 1 ([7, Theorem 10.4.6]). *Any two norms on $\mathbb{R}^n$ are equivalent.*

1.4 Inner product

An *inner product* on a vector space V is a map $\iota : V \times V \to \mathbb{R}$ with the properties
1. $\iota(u, v) = \iota(v, u)$;
2. $\iota(u, v + w) = \iota(u, v) + \iota(u, w)$;
3. $\iota(\alpha u, v) = \alpha\iota(u, v)$.

A vector space with an inner product is called an *inner product space*. The dot product is a special case of an inner product. The *dot product* of vectors is obtained in **Mathematica**® by using . or Dot. For example, the dot product of vectors in $\mathbb{R}^3$ is

> *In[·]:=* {a, b, c} . {x, y, z}
> *Out[·]:=* a x + b y + c z

We usually write $u.v$ for $\iota(u, v)$.

An inner product on a real vector space always defines a norm by the formula

$$\|v\| = \sqrt{v.v}.$$

However, not all norms arise from inner products. In fact, recall a well known fact from the Euclidean plane geometry, known as the *parallelogram law*, which says that in a parallelogram the sum of the squares of the lengths of the diagonals is equal to the sum of the squares of the lengths of the sides:

$$\|u - v\|^2 + \|u + v\|^2 = 2\|u\|^2 + 2\|v\|^2.$$

In[·]:= Block[{u = {1, 1}, v = {2, 0}}, Graphics[{Red,
    Line[{{0, 0}, u, u + v, v, {0, 0}}], Blue,
    Line[{{0, 0}, u + v}], Line[{u, v}]}]]

Out[ ]=

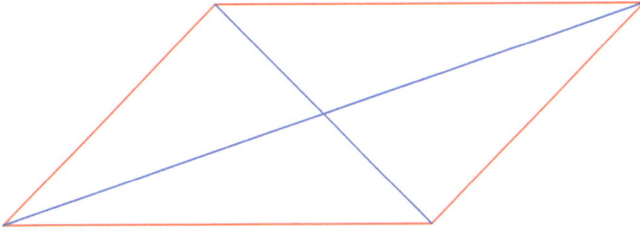

Figure 1.5

The proof is as follows. The left hand side is

$$\|u - v\|^2 + \|u + v\|^2 = (u - v).(u - v) + (u + v).(u + v).$$

In Mathematica® we have

In[·]:= TensorReduce[(u - v) . (u - v) +
    (u + v) . (u + v), Assumptions -> {Element[v,
    Vectors[n]], Element[u, Vectors[n]]}]
Out[·]:= 2 u.u + 2 v.v

which is $2\|u\|^2 + 2\|v\|^2$. The proof works in any inner product space.

Only the 2-norm arises from the inner product. We can use Mathematica® to show that neither the ∞-norm nor any p-norm for small $p \neq 2$ arises from the inner product:

In[·]:= FindInstance[2*(Norm[{x, y}, Infinity]^2 +
    Norm[{1, 0}, Infinity]^2) != Norm[{x, y} - {1, 0},
    Infinity]^2 + Norm[{x, y} + {1, 0}, Infinity]^2 &&
    Element[{x, y}, Reals], {x, y}]
Out[·]:= {{x -> -2, y -> 4}}

In the case of the standard dot product of vectors, there is a classical formula

$$u.v = \cos(\theta)\|u\| \, \|v\|,$$

where θ is the (acute) angle between the vectors u and v. In the case of the standard inner product on $\mathbb{R}^n$, one can prove that

$$|u.v| \leq \|u\| \, \|v\|,$$

which is known as the *Cauchy–Schwarz inequality*. Mathematica® can verify it for small values of n:

In[·]:= `Reduce[ForAll[{u, v, p, q}, Element[{u, v, p, q},`
`        Reals], Abs[{u, v} . {p, q}] <=`
`        Norm[{u, v}]*Norm[{p, q}]]]`
Out[·]:= `True`

1.5 Continuity of functions and mappings

We have already given the definition of *continuity of a map* between two topological spaces. A map $f : X \to Y$ is continuous if and only if for every open set $U \subset Y$ the set $f^{-1}(U)$ is open in X. When both spaces are metric spaces or normed linear spaces, the definition of continuity can also be given in the familiar Cauchy or $\varepsilon - \delta$ form, which is "constructive" and, hence, at least in certain cases, can be checked by Mathematica®.

As in the case of functions of one variable, we do not need to prove continuity of mappings by referring to the definition in every case. Instead, we proceed just as in the case of functions of one variable: we first prove that certain basic mappings are continuous and then use general theorems which state that the mappings constructed from these mappings by algebraic operations, such as addition, multiplication, etc., and compositions are continuous. The key observations are:

1. A mapping from $U \subset \mathbb{R}^n$ to $\mathbb{R}^m$ is continuous at $x \in U$ if and only if all of its components (functions from U to $\mathbb{R}$) are continuous at x.
2. The natural projections from $\mathbb{R}^n$ to $\mathbb{R}$ are continuous.
3. The maps from $\mathbb{R} \times \mathbb{R}$ to $\mathbb{R}$ given by all arithmetical operations such as $(x, y) \mapsto x + y$, $(x, y) \mapsto xy$, and $(x, y) \mapsto x^y$ (for $x \geq 0$) are continuous.

Using these properties, we can easily prove that mappings below are continuous (except possibly at the origin).

1.5.1 Examples

Let us study the continuity of the functions $f_i : \mathbb{R}^2 \to \mathbb{R}$ given by the formula

$$f_i(x, y) = \begin{cases} w_i(x, y) & (x, y) \neq (0, 0), \\ 0 & (x, y) = (0, 0), \end{cases}$$

where $w_i(x, y)$ are given by

$$w_1(x, y) = \frac{x^2 y^2}{x^2 y^2 + (x - y)^2}, \quad w_2(x, y) = \frac{x^2 y}{x^2 y^2 + x^2 + y^2}.$$

1. The function f_1 is clearly continuous at any point other than $(0, 0)$ for general reasons, so we only need to consider its behavior in a neighborhood of $(0, 0)$. If possible (i. e., in the case of functions of no more than 2 variables), it is always useful to look at the graph:

```
In[·]:= w1[x_, y_] := (x^2*y^2)/(x^2*y^2 + (x - y)^2)
In[·]:= g1 = Table[{1/n, 1/n, w1[1/n, 1/n]}, {n, 1, 10}];
In[·]:= g2 = Table[{1/n, -1/n, w1[1/n, -1/n]}, {n, 1, 10}];

In[·]:= Show[Plot3D[w1[x, y], {x, -2.4, 2.4},
        {y, -2.4, 2.4}, ColorFunction -> Function[{x, y, z},
        {Opacity[0.3], Green}], PlotRange -> All,
        Exclusions -> {{0, 0}}, Mesh -> None,
        AxesLabel -> {"x", "y", "z"}, Ticks -> None],
        Graphics3D[{Red, Point[g1]}],
        Graphics3D[{Black, Point[g2]}]]
```

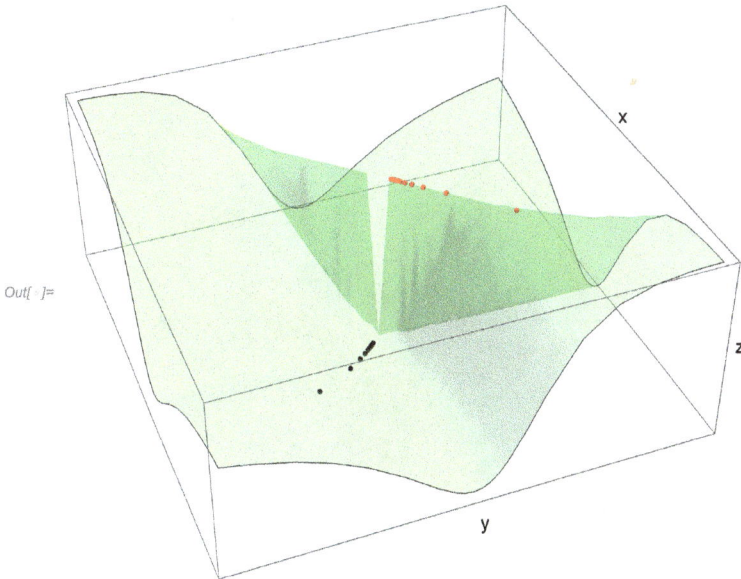

Out[·]=

Figure 1.6

Note that we used the option Exclusions to tell Mathematica® to avoid the point $(0, 0)$. In this particular case it turns out to be not necessary, but in general the option should be used for points where continuity is problematic, as Mathematica®'s graph drawing algorithm can sometimes mask a discontinuity. Here the discontinuity at $(0, 0)$ is clearly visible. We will show that f_1 is not continuous by using Heine's definition of continuity. Recall that this definition requires that for any sequence of points p_n in the domain of the function converging to $(0, 0)$, the sequence of numbers $f_1(p_n)$ must converge to $f_1(0, 0)$. However, for the sequence $(1/n, 1/n)$, which converges to $(0, 0)$ (since both components converge to 0), the sequence $f_1(1/n, 1/n)$ is constant and equal to 1, hence converges to 1. Moreover, we can see that the limit $\lim_{(x,y) \to (0,0)} f_1(x, y)$ does not even exist, as the sequence $f_1(1/n, -1/n)$ converges to 0 as n tends to infinity:

$In[\cdot]:=$ `Simplify[{w1[1/n, 1/n], w1[1/n, -1/n]}]`

$Out[\cdot]:= \left\{ 1, \dfrac{1}{1 + 4\,n^2} \right\}$

If we use the Cauchy definition, the problem can also be solved by Mathematica® alone (with the help of the Quantifier Elimination algorithm, see [5]):

$In[\cdot]:=$ `Reduce[ForAll[e, e > 0, Exists[d, d > 0,`
`        ForAll[{x, y}, {x, y} != {0, 0} && (x^2 + y^2)^(1/2)`
`        < d, Abs[w1[x, y]] < e]]]]`
$Out[\cdot]:=$ `False`

2. Let us first plot the second function f_2.

$In[\cdot]:=$ `w2[x_, y_] := (x^2*y)/(x^2*y^2 + x^2 + y^2)`
$In[\cdot]:=$ `Plot3D[w2[x, y], {x, -0.5, 0.5}, {y, -0.5, 0.5},`
`        PlotRange -> All]`

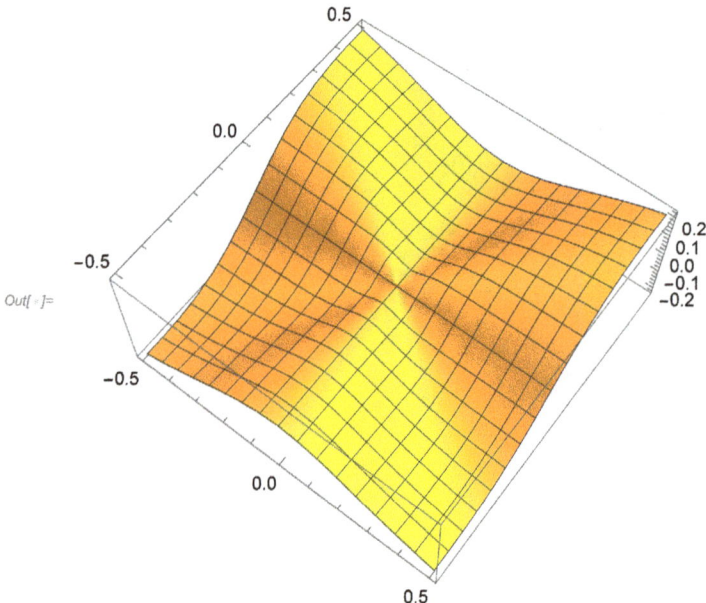

Figure 1.7

This time there are no visible discontinuities. We therefore try to prove that $f_2(x_n, y_n)$ tends to 0 whenever the sequence $(x_n, y_n) \to (0, 0)$. We have

$$|f_2(x,y)| = \left|\frac{x^2 y}{x^2 y^2 + x^2 + y^2}\right| \le \left|\frac{x^2 y}{x^2 + y^2}\right| \le |y| \left|\frac{x^2}{x^2 + y^2}\right| \le |y| \le \|(x,y)\|,$$

where $\|(x,y)\| = \sqrt{x^2 + y^2}$. If $\|(x_n, y_n)\| \to 0$, then $f_2(x_n, y_n) \to 0$.

Again, we can prove the continuity with **Mathematica**® as follows:

In[·]:= `Reduce[ForAll[e, e > 0, Exists[d, d > 0,`
`        ForAll[{x, y}, {x, y} != {0, 0} && (x^2 + y^2)^(1/2)`
`        < d, Abs[w2[x, y]] < e]]]]`

Out[·]:= `True`

3. Functions of more than two variables cannot be represented by a graph, but it is often possible to use **Mathematica**®'s function `Manipulate` to obtain an intuitive grasp of the situation. For example, consider a similar problem as above, where $f_3 : \mathbb{R}^3 \to \mathbb{R}$ is defined using the following function w_3 for non-zero (x, y, z):

In[·]:= `w3[x_, y_, z_] := (x*z + y*z)/(x^2 + y^2 + z^2)`

Although a graph of this function would require 4 dimensions, we will show its sections obtained by setting z equal to values close to 0. We will distinguish the black point p with coordinates $(z, z, 2/3)$.

In[·]:= `Manipulate[Show[Plot3D[w3[x, y, z], {x, -1, 1},`
`        {y, -1, 1}, Exclusions -> {0, 0}, Mesh -> False],`
`        Graphics3D[{Black, PointSize[0.03],`
`        Point[{z, z, 2/3}]}]], {{z, 1/2, "z"},`
`        -1, 1, Appearance -> "Labeled"}]`

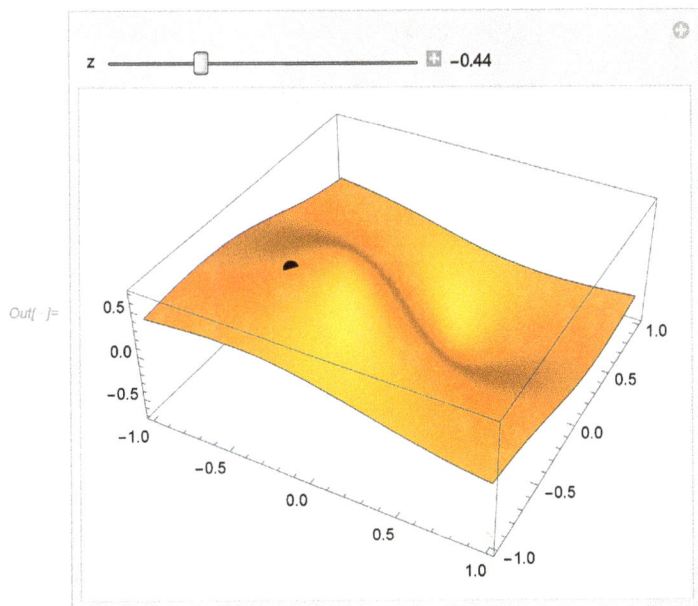

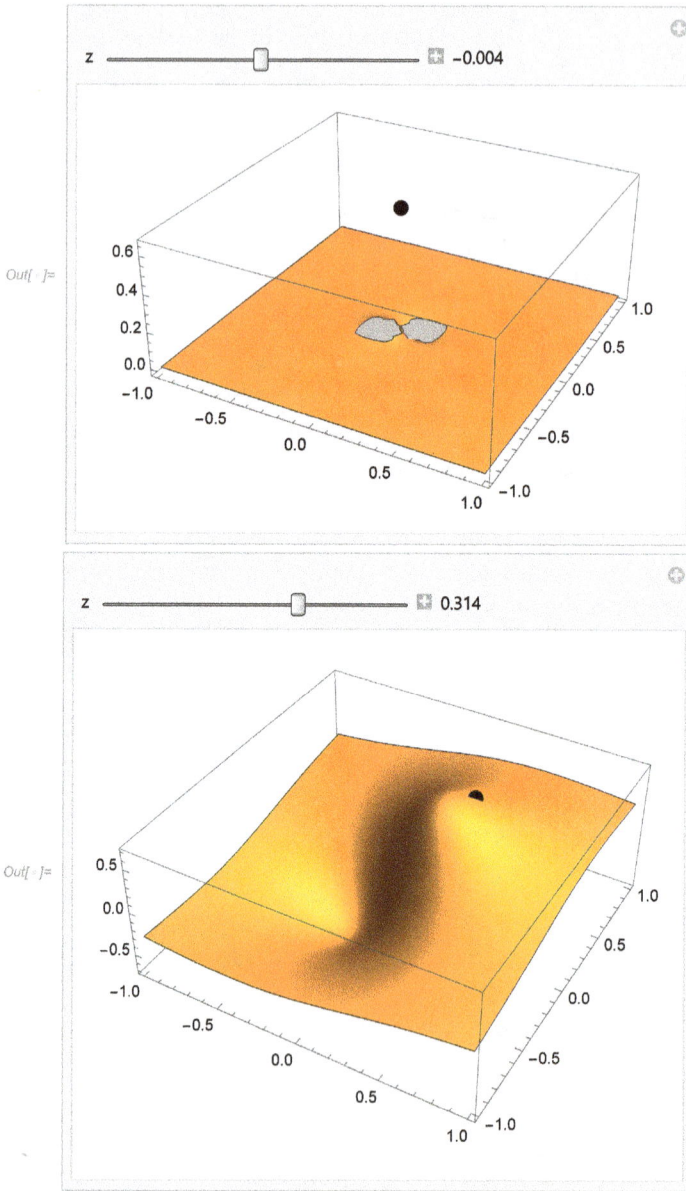

Figure 1.8

We follow the black point as $z \to 0$ and see that the point always lies on the surface (which is actually a section of the graph at a given value of z) except when we set $z = 0$. The section is then the xy-plane and the point lies above it. Thus when the point moves to $(0, 0, 0)$ along the path $x = y = z$, $f_3(p)$ moves to $2/3$ instead to 0:

In[·]:= w3[z, z, z]
Out[·]:= $\frac{2}{3}$

The function is thus discontinuous at $(0, 0, 0)$. Mathematica® can also do it quickly:

In[·]:= Reduce[ForAll[e, e > 0, Exists[d, d > 0,
        ForAll[{x, y, z}, {x, y, z} != {0, 0} &&
        (x^2 + y^2 + z^2)^(1/2) < d, Abs[w3[x, y, z]] < e]]]]
Out[·]:= False

1.6 Representing mappings graphically

Let $f : \mathbb{R}^2 \to \mathbb{R}^2$ be a mapping. The *graph of a mapping* is the set $\{(x, f(x)) \mid x \in U\}$, which cannot be represented by a picture. Sometimes we can get an idea of the behavior of the mapping by looking at the sections, as in the previous example. In the next example we use a different approach.

Consider the mapping from $\mathbb{R}^2$ to itself given by $(x, y) \mapsto (e^x \cos(y), e^x \sin(y))$. (In fact, this map is the same as the complex exponential map from $\mathbb{C}$ to $\mathbb{C}$ given by $z \mapsto \exp(z)$, where we identify the complex numbers $\mathbb{C}$ with $\mathbb{R}^2$ via the map $z \mapsto \text{Re}(z) + i \,\text{Im}(z) = x + iy$, $i^2 = -1$.) We represent the domain and the range as two parallel planes in $\mathbb{R}^3$, with the domain of the mapping above the range space. Of course, we can only display bounded subsets of both sets – in this case rectangles. We can move a point around the rectangle in the domain and view the corresponding motion of its image. For greater clarity, we have included an arrow pointing from a point in the domain to its image. The green point is the origin. For example, fixing x and varying y, we see that the image will move in a circle around the origin of radius e^x. Fixing y and increasing x, will make the image move away from 0 along a ray towards infinity, decreasing will make it move towards 0.

In[·]:= Manipulate[Module[{z = x + I*y}, Show[
        ParametricPlot3D[{x, y, 10}, {x, -20*Pi, 20*Pi},
        {y, -20*Pi, 20*Pi}], ParametricPlot3D[{x, y, 0},
        {x, -20*Pi, 20*Pi}, {y, -20*Pi, 20*Pi}],
        Graphics3D[{PointSize[0.02], Red, Point[{x, y, 10}],
        Blue, Point[{Re[Exp[z]], Im[Exp[z]], 0}],
        Green, Point[{0, 0, 0}], Red, Arrow[{{x, y, 10},
        {Re[Exp[z]], Im[Exp[z]], 0}}]}], Boxed
        -> False, Axes -> False, PlotRange -> {{-3*Pi, 3*Pi},
        {-10, 10}, {-1, 11}}]], {{x, 0, "x"}, -3*Pi, 3*Pi,
        Appearance -> "Labeled"}, {{y, 0, "y"}, -3*Pi, 3*Pi,
        Appearance -> "Labeled"}]

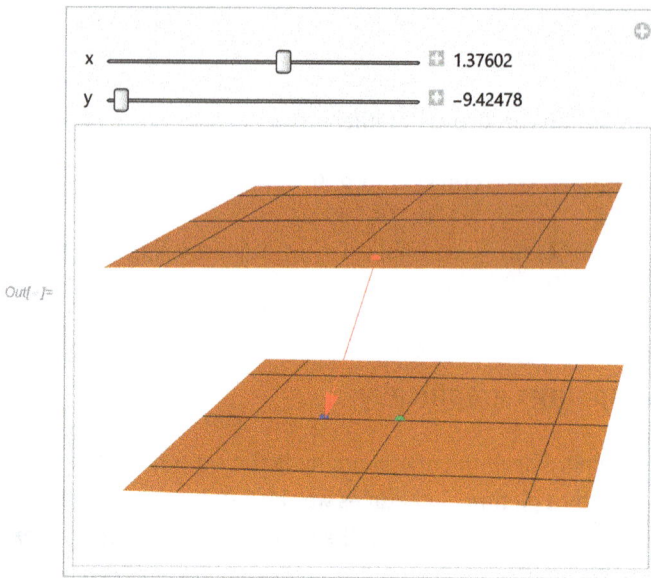

Figure 1.9

1.7 Topological properties of sets

An important topological concept in analysis is that of a *compact set*. There are several equivalent definitions of compactness of a subset of $\mathbb{R}^n$. The simplest is that a compact set (in $\mathbb{R}^n$) is one that is both closed and bounded (where bounded means that it is contained in some ball with center at the origin).

It is easy to show that a set $A \subset \mathbb{R}^n$ is a *closed set* if an only if it is sequentially closed, that is, if $a_n \to x$ as $n \to \infty$, where $a_n \in A$ for all n, then $x \in A$. It follows that compactness can also be defined sequentially: a set $A \subset \mathbb{R}^n$ is compact if and only if every sequence consisting of elements of A has a convergent subsequence with the limit in A.

For general topological spaces, compactness is defined in a different way: a set A is compact if whenever A is contained in the union of a family of open sets, it is contained in the union of some finite subfamily of this family.

We can now state a generalization of the Weierstrass theorem.

Theorem 2 ([14, Theorem 2.3.1], Continuous maps preserve compactness). *Let $A \subset \mathbb{R}^n$ be compact and suppose $f : A \to \mathbb{R}^m$ is continuous. Then the image $f(A)$ is compact.*

For $m = 1$, this implies that f is bounded and attains its supremum and infimum (see [14, [Proposition 2.3.2]).

Continuous functions can be used to study the topological properties of sets, as shown in the next example.

1.7.1 Example

Determine whether the following set

$$A = \{(x, y, z) \in \mathbb{R}^3 \mid \sin(\sin(\cos(x y))) = y \, \sin(\sin(x + y)), \; x \geq -1, \; y \leq 1\} \qquad (1.1)$$

is open, closed, bounded, or compact.

Note that since there is no restriction on z, the set is unbounded unless it is empty. So we need to find just one solution to prove that it is not. We can use FindInstance as follows:

In[·]:= `FullSimplify[FindInstance[{Sin[Sin[Cos[x*y]]]`
      `== Sin[Sin[x + y]]*y, x >= -1, y <= 1}, {x, y}]]`

Out[·]= $\left\{\left\{x \to \; \checkmark \; 89.2\ldots \, , \, y \to -818\right\}\right\}$

Figure 1.10

FindInstance found a rather complicated root, but that is all we need. Using the method of guessing, we can actually find a simpler one:

In[·]:= `Sin[Sin[Cos[x*y]]] == Sin[Sin[x + y]]*y`
      `/. {x -> 0, y -> 1}`
Out[·]:= `True`

So the set is unbounded and, hence, not compact. It is also closed because it is the intersection of the three closed sets: $p^{-1}([-1, \infty]), q^{-1}((-\infty, 1)), f^{-1}(\{0\})$, where $p(x, y, z) = x, q(x, y, z) = y, f(x, y, z) = \sin(\sin(\cos(x y))) - y \, \sin(\sin(x + y))$. Again these maps are continuous, and, hence, the inverse images of closed sets are closed.

2 Differentiation of functions of many variables

The definition of the derivative of a vector-valued function of several variables is less intuitive than for a single variable. The most obvious attempt to generalize the definition via the differential quotient does not work because division by vectors is not well defined. We will follow the usual approach of arriving at the definition in stages, keeping in mind that a satisfactory definition of differentiation must satisfy the following property: a function differentiable at a point is continuous at that point. We will begin with the simplest case of a vector-valued function of one variable.

2.1 Derivative of a vector-valued function of one variable

A continuous function $f : [a, b] \to \mathbb{R}^n$ can be thought of as describing the motion of a particle along a curve in $\mathbb{R}^n$. If at time t the particle has a well defined velocity, we say that the function is *differentiable* at t and its velocity vector is its *derivative*. As is well known from mechanics, the velocity is a vector whose components in the directions of the axes (or basic unit vectors) are the velocities of the projections of the point on the coordinate axes. Thus, the *velocity* at time t is the vector $(f_1'(t), \ldots, f_n'(t))$ of derivatives of its components. Such a function f therefore is *differentiable* if all of its components are differentiable. This, in turn, implies that they are continuous, which, as we already know, implies that the function itself is continuous.

Let us consider the familiar example. Let $f : [0, 2\pi] \to \mathbb{R}^2$ be given by $t \mapsto (\cos t, \sin t)$. The function describes a motion on the unit circle. At time t, the position vector is $(\cos t, \sin t)$ and its derivative (velocity vector) is $(-\sin t, \cos t)$. Since

In[·]:= f[t_] := {Cos[t], Sin[t]}
In[·]:= f[t] . Derivative[1][f][t]
Out[·]:= 0

this shows that the velocity vector (blue arrow on the picture below) is perpendicular to the position vector (red arrow). Note also that

In[·]:= Assuming[Element[t, Reals], Simplify[Norm[{-Sin[t],
        Cos[t]}]]]
Out[·]:= 1

so the *speed* (the magnitude of the velocity vector) is constant and the motion on the unit circle is uniform. The following dynamic picture illustrates this.

In[·]:= Manipulate[Module[{f = Function[t,
        {Cos[t], Sin[t]}], gr}, gr =
        ParametricPlot[f[s], {s, 0, 2*Pi}];
        Show[Graphics[{Red, Arrow[{{0, 0}, f[t]}],
        Blue, Arrow[{f[t], f[t] + Derivative[1][f][t]}]}],
        gr, PlotRange -> {{-2, 2}, {-2, 2}}]], {t, 0, 2*Pi}]

https://doi.org/10.1515/9783110660395-002

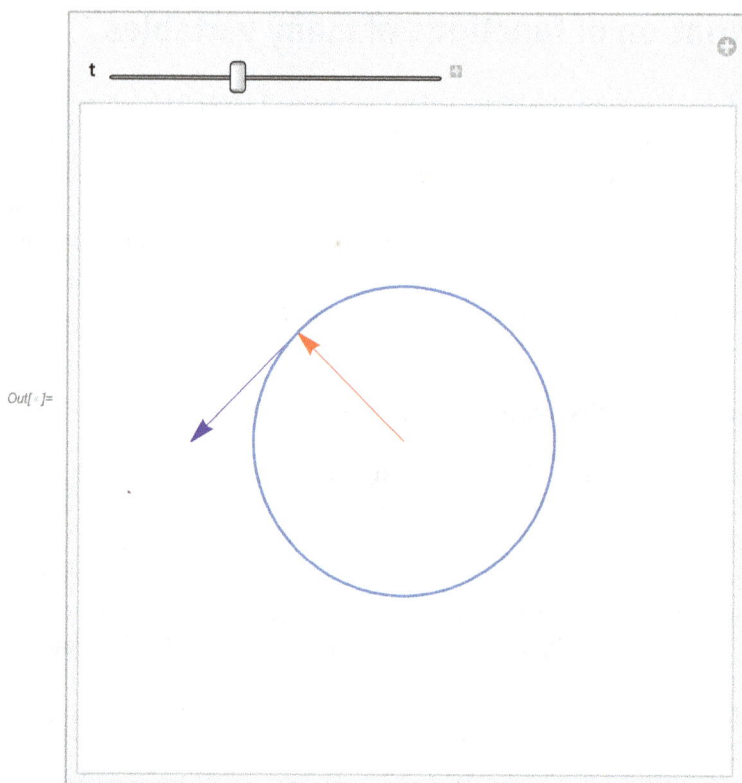

Out[ ]=

Figure 2.1

The velocity vector can be differentiated again, giving the *acceleration vector*:

In[·]:= Derivative[2][f][t]
Out[·]:= {-Cos[t], -Sin[t]}

Note that this is the opposite of the position vector, i. e., the acceleration vector has constant magnitude 1 and it points towards the center of the circle.

Of course, the motion on a circle need not be uniform. Consider now the motion given by the function

In[·]:= g[t_] := {Cos[t^2], Sin[t^2]}

A similar calculation to the above shows that the speed is no longer constant:

In[·]:= PowerExpand[Simplify[ComplexExpand[
        Norm[Derivative[1][g][t]]]]]
Out[·]:= 2 t

We can show the motion of the two particles on the same picture. The particle moving uniformly is shown in red and that whose speed is increasing is shown in green.

```
In[·]:= Manipulate[Module[{f = Function[t,
      {Cos[t], Sin[t]}], g = Function[t,
      {Cos[t^2], Sin[t^2]}], gr}, gr =
      ParametricPlot[f[s], {s, 0, 2*Pi}];
      Show[Graphics[{Red, Point[f[t]],
      Arrow[{f[t], f[t] + Derivative[1][f][t]}],
      Green, Point[g[t]], Arrow[{g[t], g[t] +
      Derivative[1][g][t]}]}], gr, PlotRange ->
      {{-10, 10}, {-10, 10}}]], {t, 0, 2*Pi}]
```

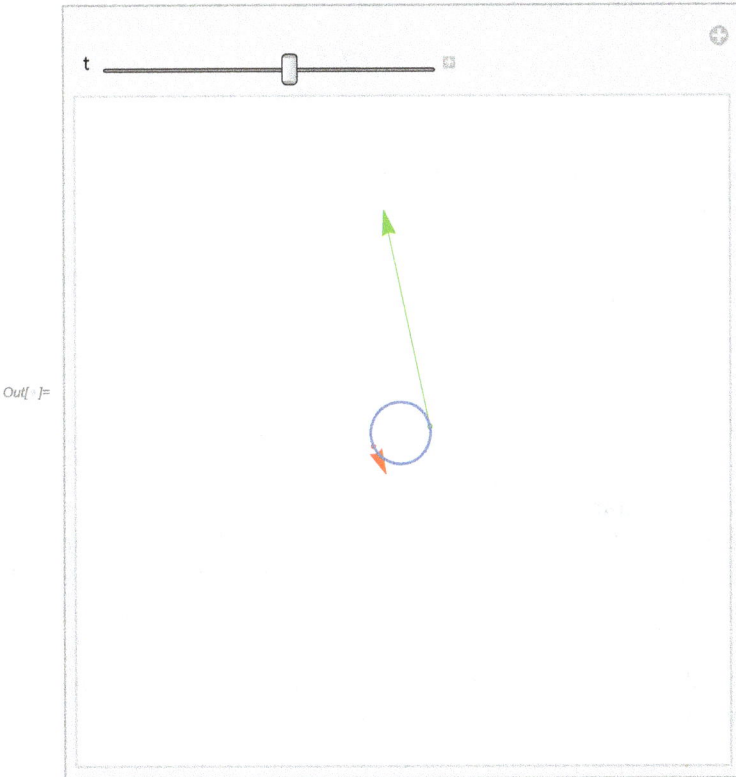

Figure 2.2

2.2 Directional and partial derivatives

Next we consider functions of several variables. Consider first a scalar function of only two variables (so that we can draw its graph):

```
In[·]:= f[x_, y_] := 2 x^2 - y^2 + x + 2 y
```

We are interested in the differentiability of this function at the point $a \in \mathbb{R}^2$ (in the illustration below a is the origin). Let us try to see if we can reduce the problem to one of computing derivatives of functions of one variable. We choose a direction in the xy-plane, that is, we choose a nonzero vector $v \in \mathbb{R}^2$ in the xy-plane represented by the black arrow in the illustration below. By restricting the function f to points lying on the straight line determined by the chosen vector, we obtain a function $f(a + tv)$ of one variable t and can ask if it is differentiable at $t = 0$. Geometrically this corresponds to the existence of a tangent vector at the origin to the curve $f(a+tv)$ which a point moving in the direction of the chosen unit vector traces on the graph of the function (see the illustration below). The *directional derivative* is thus given by the slope of this curve:

$$D_v f(a) = \lim_{t \to 0} \frac{f(a + tv) - f(a)}{t},$$

where $f : \mathbb{R}^2 \to \mathbb{R}$.

For our function, let us take as v the unit vector making the angle u with the x-axis, i. e., $v = (\cos u, \sin u)$ and $a = (0, 0)$:

In[·]:= `Limit[f[t*Cos[u], t*Sin[u]]/t, t -> 0]`
Out[·]:= `Cos[u] + 2 Sin[u]`

Note that this can also be computed as

In[·]:= `Derivative[1][Function[t, f[t*Cos[u], t*Sin[u]]]][0]`
Out[·]:= `Cos[u] + 2 Sin[u]`

The illustration below shows the graph of f, the vector v (black arrow), the path of the curve lying on the graph of f given by $t \mapsto f(a + tv)$, and its tangent vector (blue arrow) which represents the directional derivative in the direction of v. Clearly, for our example the directional derivative exists in every direction. However, as we will soon see, the existence of directional derivatives in all directions is not sufficient for differentiability. Two additional properties are needed: all the tangent vectors to the curve $f(a + tv)$ must lie on a plane (called the tangent plane, whose equation in our case is $z = 2x + y$), and this plane has to approximate the graph of the function near a.

In[·]:= `Manipulate[Module[{f = 2*#1^2 - #2^2 +`
`        #1 + 2*#2 & , g, g1 = ParametricPlot3D[`
`        {Cos[t], Sin[t], 0}, {t, 0, 2*Pi}], g2 =`
`        Plot3D[2*x + y, {x, -2, 2}, {y, -2, 2}, Mesh ->`
`        None, ColorFunction -> Function[{x, y, z},`
`        {Opacity[0.5], Yellow}]]}, g =`
`        Plot3D[f[x, y], {x, -2, 1}, {y, -2, 1}, Axes ->`
`        False, Mesh -> None, ColorFunction ->`
`        Function[{x, y, z}, {Opacity[0.5], Red}]];`
`        Show[ParametricPlot3D[{t*Cos[u], t*Sin[u],`
`        f[t*Cos[u], t*Sin[u]]}, {t, -1, 1}], Graphics3D[`

```
{Arrowheads[0.03], Blue, Arrow[{{0, 0, 0},
{Cos[u], Sin[u], Derivative[1][Function[t,
f[t*Cos[u], t*Sin[u]]]][0]}}], Black,
Arrow[{{0, 0, 0}, {Cos[u], Sin[u], 0}}]}]],
If[t, g, Graphics3D[{}]], If[s, g1, Graphics3D[{}]],
If[u1, g2, Graphics3D[{}]], Axes -> False,
PlotRange -> {{-2, 2}, {-2, 2}, {-3, 3}}]],
{u, 0, 2*Pi}, {{t, True, "show graph"},
{True, False}}, {{s, False, "show circle"},
{True, False}}, {{u1, False, "show tangent
plane "}, {True, False}}]
```

Out[ ]=

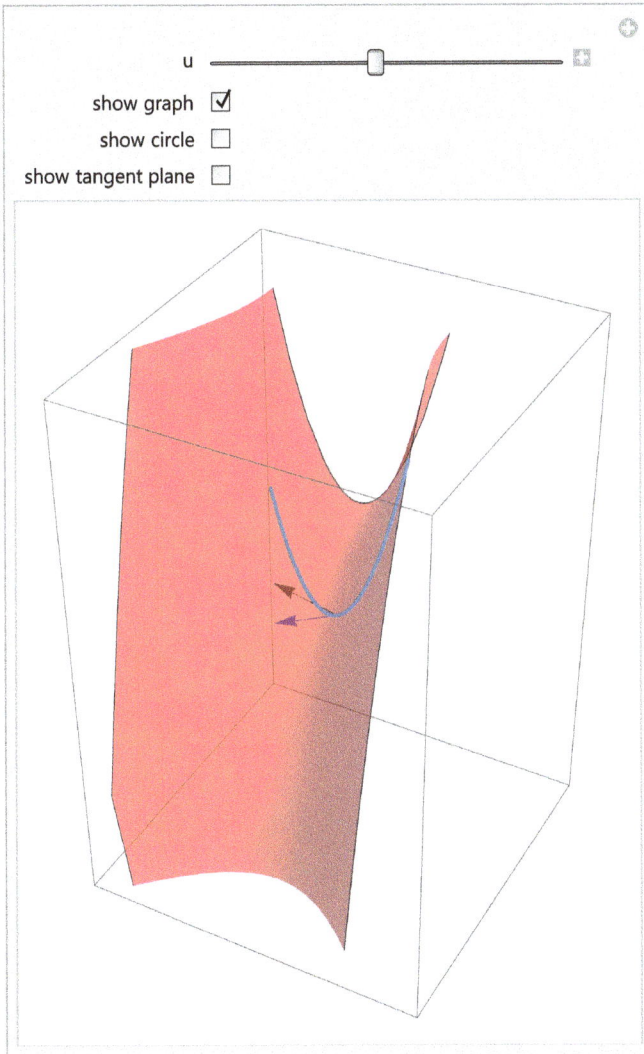

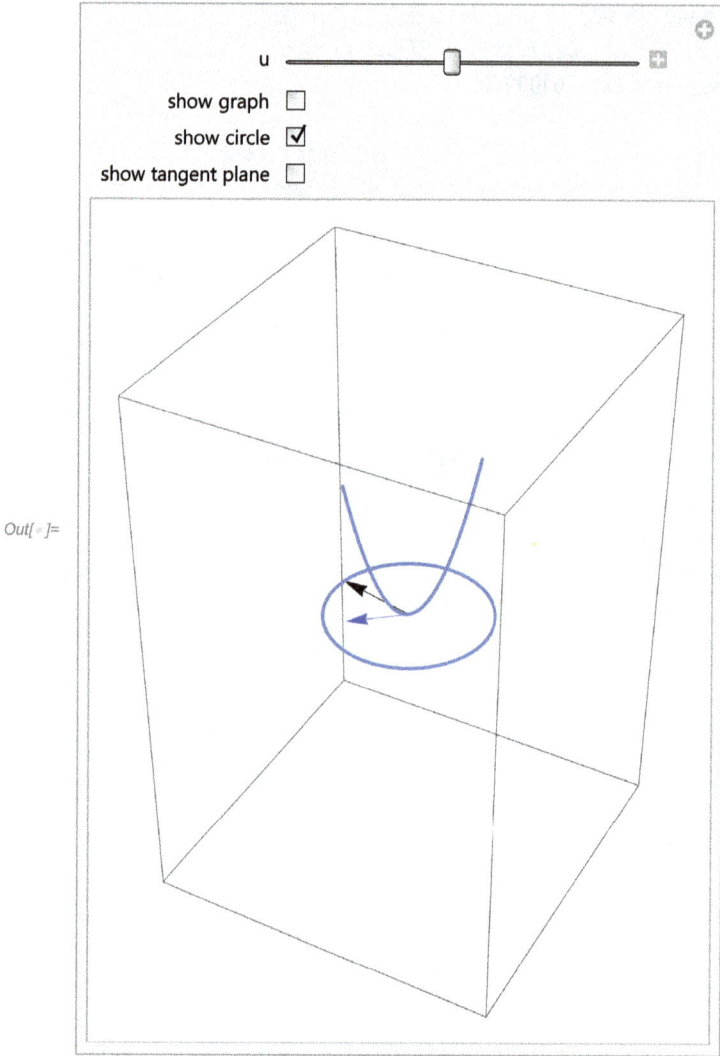

Out[ ]=

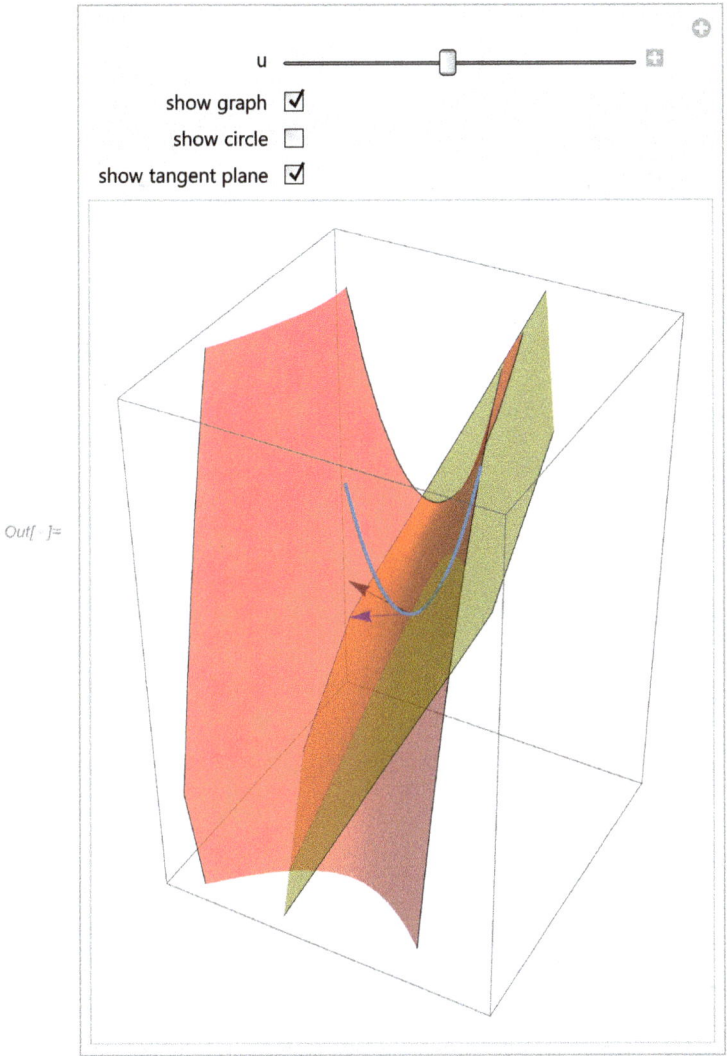

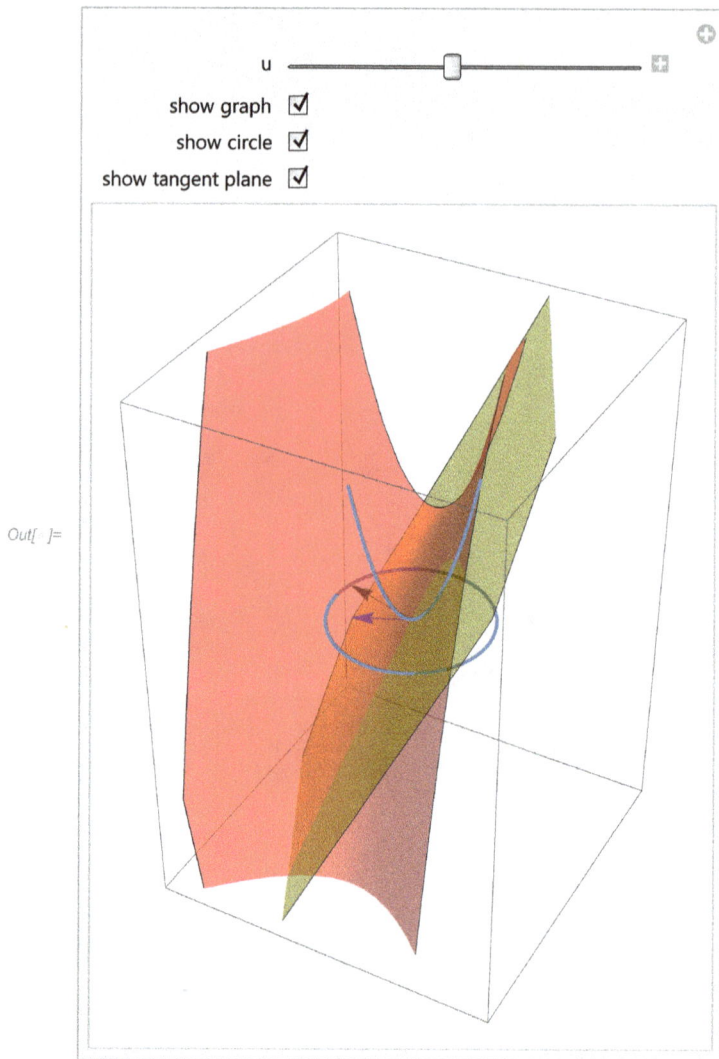

Figure 2.3

Among the directional derivatives, two are considered special — these are the derivatives in the direction of the unit vectors $(1, 0)$ and $(0, 1)$, called *partial derivatives*. They can also be defined by considering one of the variables as a constant and differentiating with respect to the other. As usual, there are two ways to compute partial derivatives in Mathematica®:

```
In[·]:= {D[f[x, y], x], D[f[x, y], y]}
Out[·]:= {1 + 4 x, 2 - 2 y}
```

In[·]:= `{Derivative[1, 0][f][x, y], Derivative[0, 1][f][x, y]}`
Out[·]:= `{1 + 4 x, 2 - 2 y}`

The quickest way to obtain all the partial derivatives of f is by using D with all the variables enclosed in double curly braces:

In[·]:= `D[f[x, y], {{x, y}}]`
Out[·]:= `{1 + 4 x, 2 - 2 y}`

The vector of all partial derivatives of a function is called the *gradient*. Mathematica® also has a built-in function Grad:

In[·]:= `Grad[f[x, y], {x, y}]`
Out[·]:= `{1 + 4 x, 2 - 2 y}`

The main difference between using Grad and using D is that Grad accepts an additional argument which represents a coordinate chart (e. g., "Spherical", see Coordinate-ChartData).

The above obviously generalizes to functions of any number of variables.

Let us now consider two examples showing that the existence of partial derivatives and even the existence of all directional derivatives is not a sufficient condition for differentiability of a function of several variables.

2.2.1 Example 1

Consider again the function we defined in the previous chapter:

$$f(x,y) = \begin{cases} \frac{x^2 y^2}{x^2 y^2 + (x-y)^2}, & (x,y) \neq (0,0), \\ 0, & (x,y) = (0,0). \end{cases}$$

We already know that the function is not continuous at $(0,0)$, therefore, it should not be differentiable there. However, clearly both partial derivatives at $(0,0)$ exist and are 0. Let us try to compute the directional derivatives at $(0,0)$ in the direction of a general vector (u,v):

In[·]:= `Cancel[((x^2*y^2)/((x - y)^2 + x^2*y^2)`
`/. {x -> t*u, y -> t*v})/t]`
Out[·]:= $\dfrac{t\,u^2\,v^2}{u^2 - 2\,u\,v + v^2 + t^2\,u^2\,v^2}$

Now we have to be a little bit careful. If we simply let Mathematica® compute this limit, we will get

In[·]:= `Limit[%, t -> 0]`
Out[·]:= `0`

which would appear to suggest that the limit exists for all vectors (u,v). However, when $u = v$, we see that the limit actually does not exist:

In[·]:= `Limit[%% /. u -> v, t -> 0]`
Out[·]:= `Indeterminate`

The reason why Mathematica® did not notice this is that, in cases involving parameters, `Limit` and many other functions consider only "generic cases", that is, cases not involving parameters having special values, or where a special relationship holds between them. Thus in this case, although both partial derivatives exist, not all directional derivatives do. In the next example, all directional derivatives exist and the function is continuous, but it still not differentiable.

2.2.2 Example 2

As we have seen in Chapter 1, the following function is continuous:

In[·]:= `Clear[f]; f[x_, y_] := Piecewise[`
`        {{(x^2*y)/(y^2*x^2 + x^2 + y^2),`
`        {x, y} != {0, 0}}}, 0]`

Let us compute its directional derivative in the direction of (u, v):

In[·]:= `Simplify[((x^2*y)/(y^2*x^2 + x^2 + y^2)`
`        /. {x -> t*u, y -> t*v})/t]`
Out[·]:= $\dfrac{u^2\,v}{v^2 + u^2\,(1 + t^2\,v^2)}$

This time there are no special cases, and the limit is

In[·]:= `Limit[%, t -> 0]`
Out[·]:= $\dfrac{u^2\,v}{u^2 + v^2}$

This limit exists for every vector (u, v) and gives the directional derivative in this direction. But unlike in Example 1, it is easy to see that the vectors $(u, v, D_{(u,v)}f(0,0))$ do not lie on a plane! (This follows from the nonlinearity of the expression for the directional derivative). Hence the function is not differentiable as there is no tangent plane to the graph at the point $(0, 0, 0)$. This fact it reflected in the "wrinkled" appearance of the graph near the origin.

In[·]:= `Plot3D[(x^2*y)/(y^2*x^2 + x^2 + y^2),`
`        {x, -1, 1}, {y, -1, 1}, ColorFunction ->`
`        Function[{x, y, z}, Hue[z]]]`

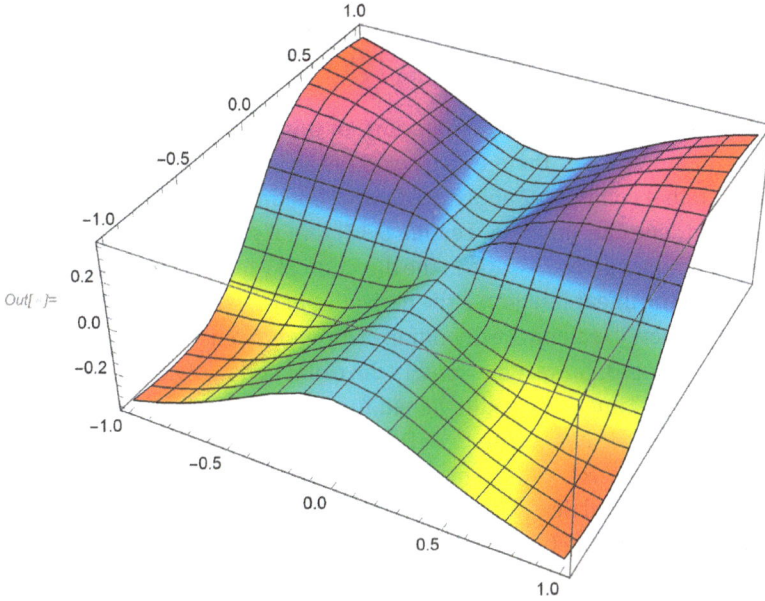

Figure 2.4

One can ask the question: suppose a function f of two variables has directional derivatives $D_{(u,v)}f(a)$ at $a \in \mathbb{R}^2$ in the direction of all vectors (u, v) and $(u, v) \mapsto D_{(u,v)}f(a)$ is linear. Will the function necessarily be differentiable? The answer is still no; it does not need even to be continuous. For example, consider the function $f : \mathbb{R}^2 \to \mathbb{R}$ which is 1 on the parabola (x, x^2) for $x \neq 0$ and 0 everywhere else. It is easy to see that all directional derivatives are 0, but the function is not even continuous. We can see that this time all the vectors $(u, v D_{(u,v)}f(a))$ lie on a plane (the xy-plane) but this plane is not tangent to the graph of the function.

2.3 Derivative of a mapping

Geometrically the concept of differentiability of a mapping $f : U \subset \mathbb{R}^2 \to \mathbb{R}$ is clear — we say that such a mapping is *differentiable* at $a = (a_1, a_2) \in U$ if the graph of f has a tangent plane at a. There are several equivalent ways to express this analytically [15, Definition 12.13, p. 711].

Definition 1. A function $f : \mathbb{R}^2 \to \mathbb{R}$ is *differentiable* at $a = (a_1, a_2)$ if
1. All directional derivatives $D_v f(a)$ exist and the function $v \mapsto D_v f(a)$ is linear;
2. One can write

$$f(a + v) - f(a) = D_v f(a) + r(a, v)\|v\|,$$

where $\lim_{\|v\| \to 0} r(a, v) = 0$.

We can, of course, replace the Euclidean norm by any other norm, since we know that they are all equivalent. In fact, instead of $\|v\| \to 0$, we can simply write $v \to 0$ (as vectors).

It is not difficult to show that the assumptions of this definition can be relaxed, and it is enough to assume that there exists some linear function $L : \mathbb{R}^2 \to \mathbb{R}$ (depending on a) such that condition 2 above holds with $L(v)$ in place of $D_v f(a)$ [14, Definition 6.2.2]. In fact, one can show [14, Lemma 6.3.5] that if such a function L exists, then $L(v)$ is the directional derivative at a in the direction of v for every nonzero vector v. It is natural to call the linear function L the derivative of f at a.

It is now obvious how to extend the definition of differentiability to mappings $U \to \mathbb{R}^m$, where $U \subset \mathbb{R}^n$. We need, however, to modify the usual notion of derivative for the cases $n = 1$ and $m = 1$. Thus, for a multi-valued function of one variable f, we will view its derivative at a as the linear map $\mathbb{R} \to \mathbb{R}^m$, $x \mapsto (f'_1(a)x, \ldots, f'_m(a)x)$, while for a map $f : U \to \mathbb{R}$, where $U \subset \mathbb{R}^n$, it is the map $\mathbb{R}^n \to \mathbb{R}$, $(v_1, \ldots, v_n) \mapsto \sum_{i=1}^n v_i \partial f(x)/\partial x_i$, where $\partial f(x)/\partial x_i$ are the partial derivatives. Note that sometimes we shall use notation $'$ for a derivative of the mapping (we should not forget that $f'(x)$ now denotes a linear transformation, not a number).

2.3.1 Example 3

Let us investigate the differentiability of the function $f : \mathbb{R}^3 \to \mathbb{R}$ given by $f(x,y,z) = xyz + x^2$.

There are several methods of solving this problem. Let us start by applying the definition directly.

```
In[·]:= f[x_, y_, z_] := x*y*z + x^2
```

```
In[·]:= h[x_, y_, z_] = Expand[Simplify[
          f[x + u, y + v, z + w] - f[x, y, z]]]
Out[·]:= u² +u v w + 2 u x + v w x + u w y +
         w x y + u v z + v x z + u y z
```

We form the list of monomials and select those linear in u, v, w:

```
In[·]:= l[u_, v_, w_] = Plus @@ Select[MonomialList[
          h[x, y, z], {u, v, w}], Total[Exponent[#,
          {u, v, w}]] <= 1 &]
Out[·]:= w x y + v x z + u (2 x + y z)
```

This will give us the derivative at the point (x, y, z) provided we prove that the remainder satisfies the necessary condition. The remainder is

In[·]:= `ExpandAll[(h[x, y, z] - l[u, v, w])/`
`        Sqrt[u^2 + v^2 + w^2]]`

Out[·]:= $\dfrac{u^2}{\sqrt{u^2 + v^2 + w^2}} + \dfrac{u\,v\,w}{\sqrt{u^2 + v^2 + w^2}} + \dfrac{v\,w\,x}{\sqrt{u^2 + v^2 + w^2}}$

$+ \dfrac{u\,w\,y}{\sqrt{u^2 + v^2 + w^2}} + \dfrac{u\,v\,z}{\sqrt{u^2 + v^2 + w^2}}$

Now suppose that u, v, w all tend to 0. It is then easy to see that all the terms in the above expression tend to 0. For example,

$$\left| \frac{u^2}{\sqrt{u^2 + v^2 + w^2}} \right| = |u| \frac{|u|}{\sqrt{u^2 + v^2 + w^2}} \le |u|.$$

Hence it tends to 0 as $|u|$ tends to 0. We have not only shown that the function is differentiable but also found the derivative.

2.4 Functions of class C^1

Checking the differentiability of a mapping by using the definition is usually inconvenient, so we will describe another approach. Before that we need to say more about partial derivatives. As we have already mentioned, given a function of n variables $f : U \to \mathbb{R}$ and a point $a \in U \subset \mathbb{R}^n$, its partial derivatives $\partial f(x)/\partial x_i$ are the directional derivatives in the direction of the basis vectors $e_i = (0, \ldots, 1, \ldots, 0)$. They can also be computed as ordinary derivatives of the functions $x \mapsto f(a_1, \ldots, x, \ldots, a_n)$ at $x = a_i$. If all the partial derivatives of f exist at all the points of U and are continuous functions on U, we say that f is of *class* C^1. We have the following theorem (see [15, Theorem 12.21] for a statement with weaker assumptions).

Theorem 3. *Let $f : U \to \mathbb{R}$, where U is open in $\mathbb{R}^n$. Suppose that all the partial derivatives are defined and continuous on U. Then f is differentiable at all points of U.*

This, of course, immediately solves the problem in Example 3 above, since the partial derivatives of a polynomial are themselves polynomials and thus are continuous. Of course, in this particular case an even simpler approach can be used which is similar to that we have already used for continuity: we prove the various "arithmetical properties" of differentiability, i. e., that the sums, products, and quotients of differentiable functions are differentiable. Also the standard projections of $\mathbb{R}^n$ to $\mathbb{R}^m$, where $n > m$, are easily seen to be differentiable. This immediately implies that all polynomials are everywhere differentiable.

2.4.1 Example 4

Let us consider a more difficult example. Is the following function:

$$f(x,y) = \begin{cases} \frac{x^2y^2}{x^2+y^2}, & (x,y) \neq (0,0), \\ 0, & (x,y) = (0,0) \end{cases}$$

differentiable?

We already know that it is differentiable at every point except perhaps $a = (0,0)$. We can again use two approaches. First, we use the definition of differentiability. Here we have to use the fact (easily proved) that for a function $f : \mathbb{R}^n \to \mathbb{R}$ the decomposition

$$f(a+v) = f(a) + L(v) + r(a,v)\|v\|,$$

where L is linear and $r(a,v) \to 0$ as $v \to 0$, is unique (if such a decomposition exists, see [14, Lemma 6.2.4]). In our case L is the zero function, provided we can show that $x^2y^2/(x^2+y^2)^{3/2} \to 0$ as $x^2 + y^2 \to 0$. We have

$$\left| \frac{x^2y^2}{(x^2+y^2)^{3/2}} \right| = \frac{x^2}{x^2+y^2} \frac{|y|}{\sqrt{x^2+y^2}} |y| \le |y| \to 0.$$

So the function is differentiable.

Alternatively, we can use the above theorem. First we note that the partial derivatives at $(0,0)$ exist and are both equal to 0. Then we compute the partial derivatives at other points

$$In[\cdot] := \texttt{Simplify[D[(x\^2*y\^2)/(x\^2 + y\^2), \{\{x, y\}\}]]}$$

$$Out[\cdot] := \left\{ \frac{2\,x\,y^4}{(x^2+y^2)^2}, \frac{2\,x^4\,y}{(x^2+y^2)^2} \right\}$$

Now we need to show that these functions are continuous at $(0,0)$. For example, for the first partial derivative we have

$$\left| \frac{2xy^4}{(x^2+y^2)^2} \right| = 2|x| \frac{y^4}{(x^2+y^2)^2} \le 2|x| \to 0$$

as $x \to 0$. For the second partial derivative, the computations are similar.

2.5 The Jacobian matrix

We have seen that a scalar-valued function of two variables is differentiable at a point if its graph has a tangent plane at that point. This is equivalent to saying that the function can be approximated by a linear function in a neighborhood of the point.

For a mapping $f : U \subset \mathbb{R}^n \to \mathbb{R}^m$, the matrix representing its derivative is called the *Jacobian matrix* which we shall denote by J. It is the matrix of partial derivatives [14, Section 6.3]. If we know J for a differentiable mapping, then we can obtain the directional derivative in the direction of a vector v by computing Jv, where v is the column vector. Note that in the case of scalar-valued functions the Jacobian matrix is usually replaced by a vector – the gradient, which can also be viewed as a linear map acting on vectors by the dot product.

As we have already seen, the Jacobian matrix may exist even when the function is not differentiable, but in that case it does not represent the derivative. Let us now consider how to compute the Jacobian matrix with Mathematica®. Consider a map from $\mathbb{R}^2$ to itself given by

In[·]:= f[x_, y_] := {x^2 + x y, y^2}

We can compute its Jacobian matrix with Mathematica® as follows:

In[·]:= D[f[x, y], {{x, y}}]
Out[·]:= {{2 x + y, x}, {0, 2 y}}

Now recall that matrices represent linear transformations and act on vectors by matrix multiplication, where we view a vector as a column matrix (usually thought of as a "column vector"). In other words, a matrix $(a_{i,j})_{i,j=1}^2$ acts on a vector $(x, y)^{tr}$ by

In[·]:= MatrixForm[{{a11, a12}, {a21, a22}} . {{x}, {y}}]
Out[·]//MatrixrForm= $\begin{pmatrix} a11\,x + a12\,y \\ a21\,x + a22\,y \end{pmatrix}$

However, since writing column matrices in Mathematica® is tedious, its easier to use Mathematica®'s feature that allows matrices to act on ordinary row vectors as follows:

In[·]:= {{a11, a12}, {a21, a22}} . {x, y}
Out[·]:= {a11 x + a12 y, a21 x + a22 y}

Note that the result is another (row) vector.

In fact, Mathematica®'s Dot product (usually denoted by .) is defined more generally for multiplying tensors (generalizations of vectors and matrices) of subtable shape. However, the only cases we will use are the dot product of two vectors, the dot product of a matrix and a vector (operation of a linear transformation on a vector), and the dot product (multiplication) of matrices (composition of linear transformations).

2.5.1 Example 5

Consider the mapping $f : \mathbb{R}^2 \to \mathbb{R}^2$ given by

In[·]:= f[x_, y_] := {x + y^2, x^2 + y}

The mapping is differentiable with the Jacobian matrix

In[·]:= D[f[x, y], {{x, y}}]
Out[·]:= {{1, 2 y}, {2 x, 1}}

The directional derivative at the point $(0, 1)$ in the direction of vector (u, v) is

In[·]:= (D[f[x, y], {{x, y}}] /. {x -> 0, y -> 1}).{u, v}
Out[·]:= {u + 2 v, v}

2.5.2 Example 6

As we mentioned before, when the function is not differentiable, multiplying vectors by the Jacobian matrix does not necessarily give the directional derivatives. For example, consider again the function $f : \mathbb{R}^2 \to \mathbb{R}$ given by

In[·]:= f[x_, y_] := Piecewise[{{(x^2*y)/(y^2*x^2 +
            x^2 + y^2), {x, y} != {0, 0}}}, 0]

For a scalar function, we can use the gradient in place of the Jacobian matrix. The gradient is clearly zero but, as we saw in Example 2, the directional derivative in the direction $(1, 1)$ is 1/2.

2.5.3 Example 7

Let $M_{n \times m}$ denote the vector space of real $n \times m$ matrices. Compute the derivative of the mapping $f : M_{n \times n} \to M_{n \times n}$, where $f(X) = X^2$.

Let us first look at the case $n = 2$. We can then write our map explicitly as

In[·]:= {{x, y}, {u, v}} -> MatrixPower[{{x, y},
            {u, v}}, 2]
Out[·]:= {{x, y}, {u, v}} → {{x² + u y, v y + x y},
            {u v + u x, v² + u y}}

This mapping can be identified with the following mapping from $\mathbb{R}^4$ to itself:

In[·]:= Flatten[{{x, y}, {u, v}}] ->
            Flatten[MatrixPower[{{x, y}, {u, v}}, 2]]
Out[·]:= {x, y, u, v} → {x² + u y, v y + x y, u v + u x, v² + u y}

We can now compute the Jacobian matrix:

In[·]:= J = D[{u*y + x^2, v*y + x*y, u*v + u*x,
            u*y + v^2}, {{x, y, u, v}}]
Out[·]:= {{2 x, u, y, 0}, {y, v + x, 0, y}, {u, 0, v + x, u},
            {0, u, y, 2 v}}

For a general n, we can see that the all the components of the map are polynomial functions, hence the partial derivatives are continuous. Hence the mapping is differentiable, and to find its derivative we only need to find its directional derivatives. We have

$$\frac{f(X + tY) - f(X)}{t} = \frac{(X + tY)(X + tY) - X^2}{t}$$
$$= \frac{X^2 + tXY + tYX + t^2Y^2 - X^2}{t} = XY + YX + tY^2.$$

(Recall, that multiplication of matrices is non-commutative). Taking the limit as $t \to 0$, we see that the derivative at X is given by $XY + YX$.

Let us check that this agrees with the answer we obtained for the case $n = 2$ by direct calculation. From the first answer, we have

```
In[·]:= J . {a, b, c, d}
Out[·]:= {b u + 2 a x + c y, b (v + x) + a y + d y,
          a u + d u + c (v + x), b u + 2 d v + c y}
```

We can check that it is equal to $XY + YX$ with

$$X = \begin{pmatrix} x & y \\ u & v \end{pmatrix}, \quad Y = \begin{pmatrix} a & b \\ c & d \end{pmatrix}.$$

Indeed, we have

```
In[·]:= Flatten[{{x, y}, {u, v}} . {{a, b}, {c, d}}
        + {{a, b}, {c, d}} . {{x, y}, {u, v}}]
Out[·]:= {b u + 2 a x + c y, b v + b x + a y + d y,
          a u + d u + c v + c x, b u + 2 d v + c y}
```

2.6 The chain rule

The *chain rule* for one variable tells us how to differentiate the composition of two functions, and it is applied by Mathematica® automatically:

```
In[·]:= Derivative[1][f @* g]
Out[·]:= f'[g[#1]] g'[#1] &
```

or

```
In[·]:= D[f[g[x]], x]
Out[·]:= f'[g[x]] g'[x]
```

Mathematica® also knows the analogue of this for mappings. For example, consider the composition of two mappings $f, g : \mathbb{R}^2 \to \mathbb{R}^2$, where $f(x, y) = (f_1(x, y), f_2(x, y))$ and $g(x, y) = (g_1(x, y), g_2(x, y))$. Then we can show that

```
In[·]:= Simplify[D[{f1[g1[x, y], g2[x, y]],
        f2[g1[x, y], g2[x, y]]}, {{x, y}}] ==
        (D[{f1[u, v], f2[u, v]}, {{u, v}}] /. {u -> g1[x, y],
        v -> g2[x, y]}) . D[{g1[x, y], g2[x, y]}, {{x, y}}]]
Out[·]:= True
```

This is a special case of a general theorem (see [14, Theorem 6.4.1]).

Theorem 4. *Let U be a subset of $\mathbb{R}^n$ and let V be a subset of $\mathbb{R}^m$. Let g : U → V be a mapping, and let f : V → $\mathbb{R}^p$ be another mapping. Let x_0 be a point in the interior of U. Suppose that g is differentiable at x_0 and that $g(x_0)$ is in the interior of V. Suppose also that f is differentiable at $g(x_0)$. Then f ∘ g : U → $\mathbb{R}^p$ is also differentiable at x_0, and we have the formula*

$$(f \circ g)'(x_0) = f'(g(x_0)) \circ g'(x_0).$$

Here ∘ denotes the composition of mappings. Since all the derivatives are linear mappings, they can be represented by matrices and, under this representation, their composition corresponds to multiplication of matrices. The theorem can therefore be written in matrix form:

$$J(f \circ g)(x_0) = J(f)(g(x_0))J(g)(x_0),$$

where $J(f)(a)$ denotes the Jacobian matrix of f at a point a.

2.7 Level sets of differentiable functions

We will now consider an important geometric application. Consider a function $f : U \subset \mathbb{R}^n \to \mathbb{R}$. By the *level set* of such a function corresponding to a value $b = f(a) \in \mathbb{R}$, where $a \in U$, we mean the preimage $f^{-1}(b)$. When $n = 2$, we also call the level set a level curve, and when $n = 3$, a level surface. For arbitrary n, the level set is also called a hypersurface.

In Mathematica® a level curve (or curves) can be obtained by means of the function ContourPlot and level surfaces by means of ContourPlot3D (level sets are also called contours). Consider, for example, the function defined on $\mathbb{R}^3$ given by

```
In[·]:= f[x_, y_, z_] := x^4 + y^4 + z^2 - 1
```

To plot the level surface $f^{-1}(0)$, we use

```
In[·]:= surf = ContourPlot3D[f[x, y, z] ==
        0, {x, -2, 2}, {y, -2, 2}, {z, -2, 2},
        Mesh -> False, ColorFunction ->
        Function[{x, y, z}, Opacity[0.5, Blue]]]
```

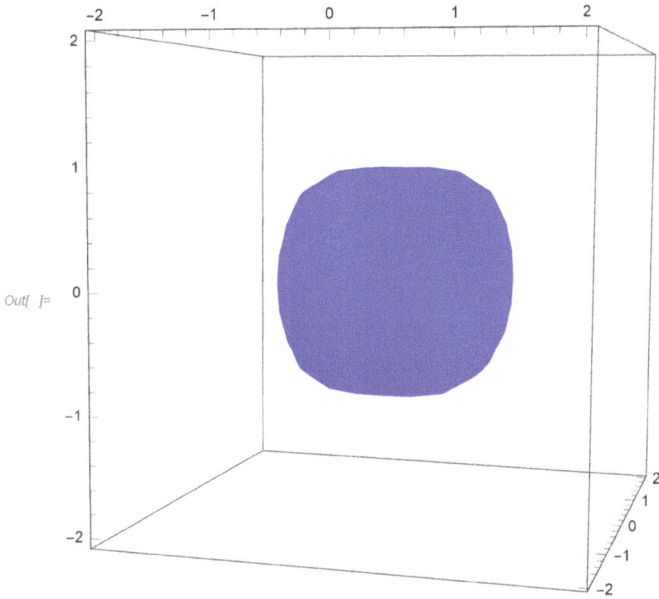

Figure 2.5

We can also see several level surfaces at once, e. g.,

```
In[·]:= ContourPlot3D[f[x, y, z], {x, -2, 2},
         {y, -2, 2}, {z, -2, 2}, Contours -> 10]
```

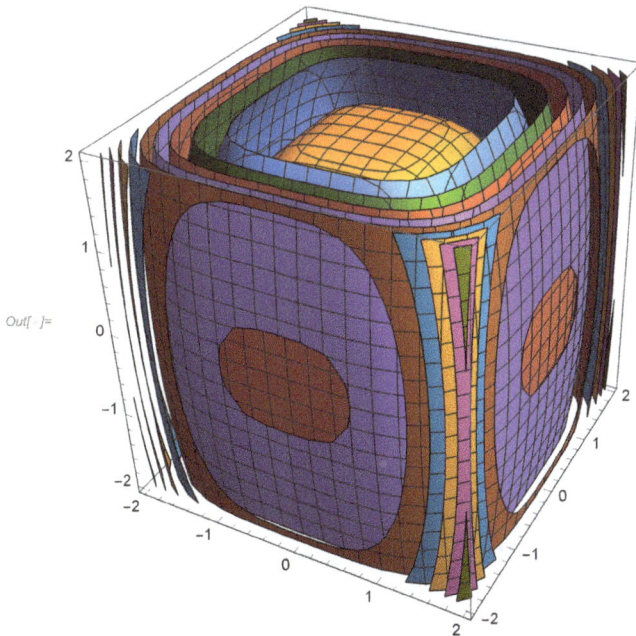

Figure 2.6

Recall that a continuous curve in $\mathbb{R}^n$ is a continuous mapping $g : [a, b] \to \mathbb{R}^n$. We say that the curve lies on a level surface $f(x) = 0$ if for each $t \in [a, b]$ we have $f(g(t)) = 0$. For example, $g : [0, \pi/2] \to \mathbb{R}^3$ lies on the level surface above:

```
In[·]:= g[t_] := {Sqrt[Cos[t]]*Sqrt[Sin[t]],
         Sqrt[Sin[t]]*Sqrt[Sin[t]], Cos[t]}
```

```
In[·]:= Simplify[f[g[t][[1]], g[t][[2]], g[t][[3]]]]
Out[·]:= 0
```

```
In[·]:= curve = ParametricPlot3D[g[t], {t, 0, Pi/2},
         ColorFunction -> Function[{x, y, z}, Red]];
```

```
In[·]:= Show[surf, curve]
```

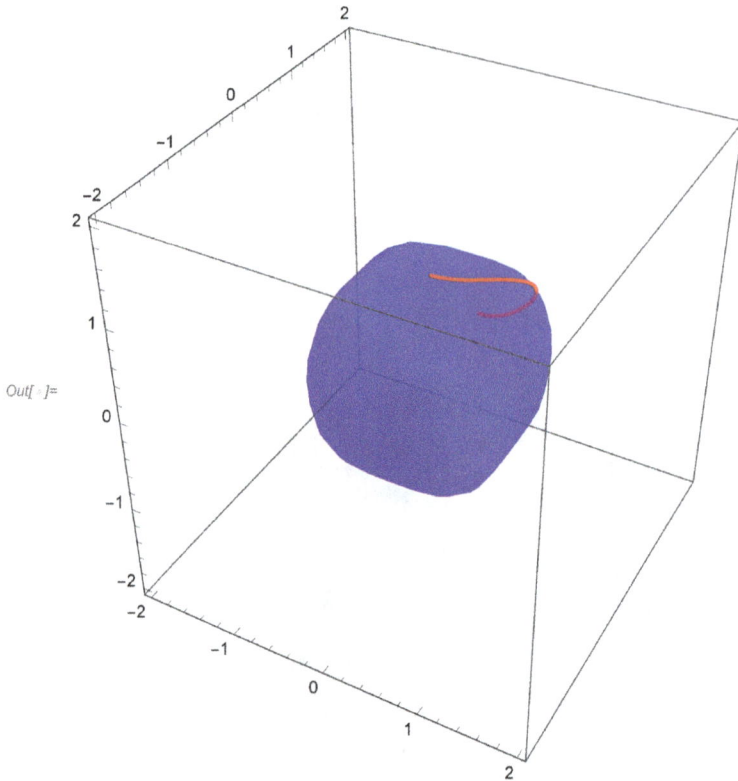

Out[·]=

Figure 2.7

We already know that the derivative of such a differentiable mapping defines a vector field tangent to the curve at every point on the curve. (By a *vector field* on a subset $U \subset \mathbb{R}^n$ we mean a vector-valued function). The gradient of our function f is a vector field defined on the entire $\mathbb{R}^3$. We will now show that the gradient is perpendicular (normal) at every point to a tangent at the same point to any smooth curve (or the so-called *regular curve* with nowhere vanishing tangent vector field) passing through this point and lying on the level surface. Let us compute the gradient at the points of our curve *g*:

```
In[·]:= nr[u_] = Grad[f[x, y, z], {x, y, z}]
          /. Thread[{x, y, z} -> g[u]];
```

In the dynamic illustration below we show both the tangent and gradient vectors on the curve.

```
In[·]:= Manipulate[Show[{surf, curve,
          Graphics3D[{Red, Arrow[{g[u], g[u] +
          Normalize[Derivative[1][g][u]]}],
          Purple, Arrow[{g[u], g[u] +
          Normalize[nr[u]]}]}]}], {u,
          0.01, Pi/2}, SaveDefinitions -> True]
```

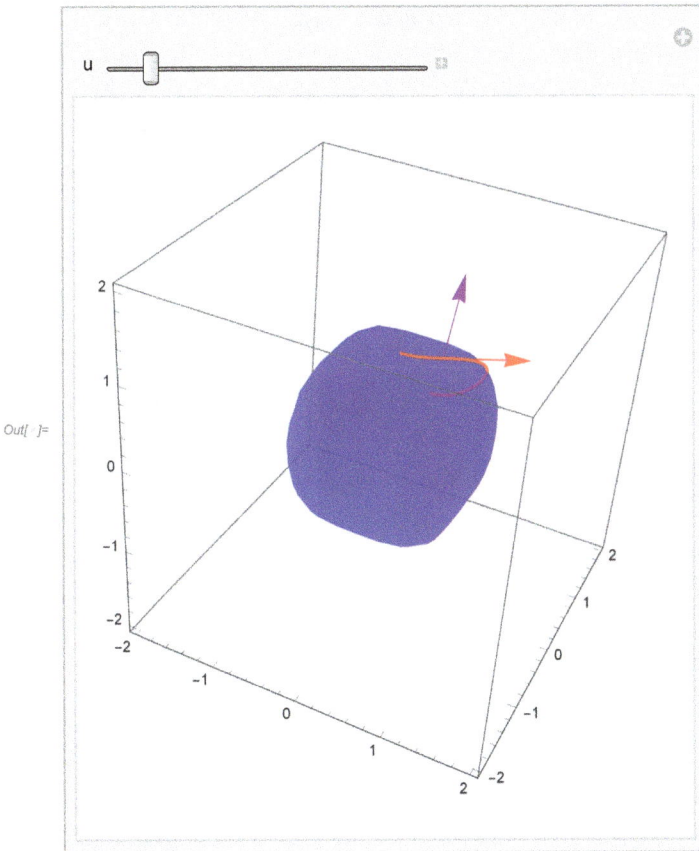

Figure 2.8

We see that these vectors are perpendicular:

> *In[·]:=* Simplify[nr[u] . Derivative[1][g][u]]
> *Out[·]:=* 0

Let us now state and prove the general theorem.

Theorem 5. *Let $U \subset \mathbb{R}^n$ be an open set, let $f : U \to \mathbb{R}$ be a differentiable function, and let $x_0 \in U$. Then $\nabla f(x)$ is normal at x to the level set $L = f^{-1}(f(x_0))$ for every $x \in L$.*

To prove the theorem, we consider any smooth (that is, continuous and differentiable in the interval) curve $g : [a, b] \to L \subset U$ such that $g(\tau) = x$ for some $\tau \in (a, b)$. Let $h : [a, b] \to \mathbb{R}$ be the composite function of $g : [a, b] \to L$ and $f : U \to \mathbb{R}$. Since the image of g lies in L, h is constant ($h(t) = f(x_0)$ for all $t \in [a, b]$). Hence $h'(t) = 0$ and, in particular, $h'(\tau) = 0$. Now applying the chain rule, we get $(\nabla f(x)) \cdot g'(\tau) = 0$, which means that the gradient $\nabla f(x)$ is perpendicular to the tangent vector $g'(\tau)$.

It is possible to prove a stronger result: the vector space of vectors in $\mathbb{R}^n$ perpendicular to $\nabla f(x)$ coincides with the tangent space to a hyperspace L at $x \in L$.

Note that this theorem tells us how to find the *tangent plane* to a level set of a differentiable function at a given point. We illustrate it on our example above. Let $p = (1/\sqrt{2}, 1/\sqrt{2}, 1/\sqrt{2})$, which clearly satisfies $x^4 + y^4 + z^2 = 1$. The gradient at p is

> *In[·]:=* p = {1/Sqrt[2], 1/Sqrt[2], 1/Sqrt[2]};

> *In[·]:=* v = Grad[x^4 + y^4 + z^2, {x, y, z}] /.
>   Thread[{x, y, z} -> p]
> *Out[·]:=* {$\sqrt{2}$, $\sqrt{2}$, $\sqrt{2}$}

The equation of the tangent plane is

> *In[·]:=* eq = Simplify[({x, y, z} - p) . v == 0]
> *Out[·]:=* $\sqrt{2}$ (x + y + z) == 3

The illustration below shows our surface and the tangent plane at p.

> *In[·]:=* tg = Plot3D[Evaluate[z /. Solve[eq, z][[1]]],
>   {x, 0, 2}, {y, 0, 2}];

> *In[·]:=* Show[surf, tg]

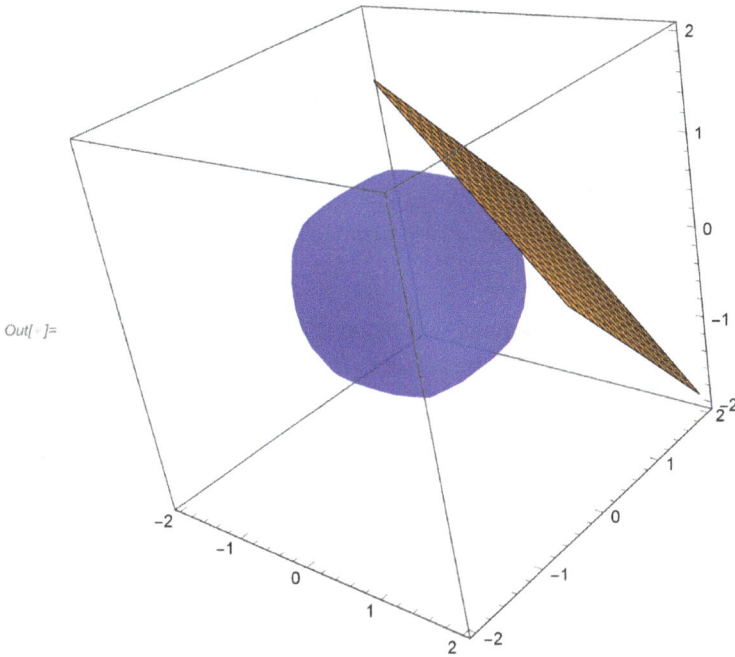

Out[ ]=

Figure 2.9

2.8 Maxima and minima

For a differentiable function $f : U \subset \mathbb{R}^n \to \mathbb{R}$, where U is an open set, a point $x_0 \in U$ is called a *critical point* if $\nabla f(x_0) = 0$, i. e., if all the partial derivatives of f at x_0 vanish. Just as in the one variable case, a necessary condition for f to have a local extremum at x_0 is that x_0 is a critical point of f (a statement similar to Fermat's lemma). Although this lemma gives us only a necessary condition for an extremum, in certain situations it is sufficient to determine it.

Let us consider an example. Suppose that the function $f : \mathbb{R}^2 \to \mathbb{R}$ is given by

In[·]:= `Clear[f]; f[x_, y_] := x*y*(1 - x)*(2 - y) + 2`

We will first consider the restrictions of this function to closed disks $B(r)$ centered at the origin and find the maximum and minimum on each of them. Let us start by finding all the critical points of the function in the interior of the disk $B(1)$. For this purpose, we need to solve the system:

In[·]:= `reg = {x, y} /. Solve[D[f[x, y], {{x, y}}]`
`       == 0 && x^2 + y^2 < 1, {x, y}]`
Out[·]:= `{{0, 0}}`

In[·]:= f[0, 0]
Out[·]:= 2

So $(0,0)$ is the only critical point in the interior of the compact set $B(1)$. By the Weierstrass theorem, we know that both the global supremum and global infimum are attained either in the interior or on the boundary of this set. So now we need only to consider the restriction of f to the boundary. We can parametrize it by $x = \cos u$, $y = \sin u$ as u runs from 0 to 2π. We get a function of one variable

In[·]:= g[u_] := f[Cos[u], Sin[u]]

Looking at the plot, we see that extrema will be on the boundary. We can use the method of Volume I [5] to actually find the values.

In[·]:= Plot[g[u], {u, 0, 2*Pi}]

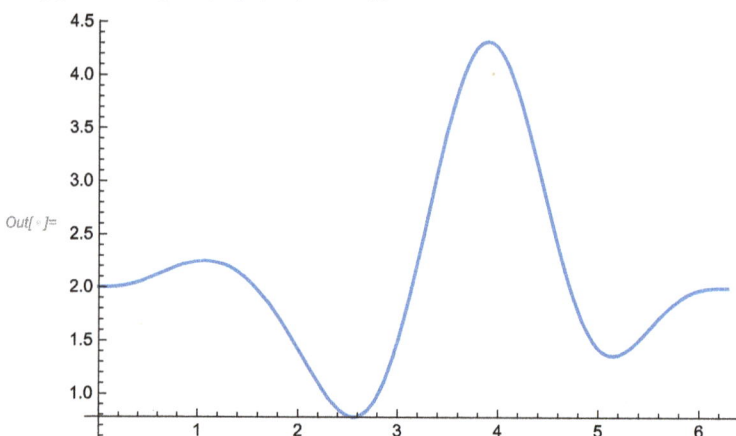

Out[·]=

Figure 2.10

Mathematica® can easily solve this problem:

In[·]:= N[Maximize[{f[x, y], x^2 + y^2 <= 1}, {x, y}]]
Out[·]:= {4.31617, {x -> -0.728768, y -> -0.684761}}

In[·]:= N[Minimize[{f[x, y], x^2 + y^2 <= 1}, {x, y}]]
Out[·]:= {0.777671, {x -> -0.836811, y -> 0.547492}}

We can now enlarge the radius of the disc and use the same procedure. As we increase the size of the disc, the number of critical points in the interior will increase until the maximum number of 5 is reached. In this particular example Mathematica® can find all 5 critical points:

In[·]:= Solve[D[f[x, y], {{x, y}}] == 0, {x, y}]
Out[·]:= $\left\{ \{x \to 0, y \to 0\}, \{x \to 1, y \to 0\}, \{x \to \frac{1}{2}, y \to 1\}, \right.$
$\left. \{x \to 0, y \to 2\}, \{x \to 1, y \to 2\} \right\}$

In the dynamic illustration below, we show the graph of the function, the boundary circle, and critical points (red) with critical values (yellow). We can also observe how the function changes along the boundary by checking the checkbox "points on the boundary". Note that vertical line shown should connect the boundary circle with the graph of the function. However, when the value of the function is too large, the point where the line meets the graph of the function will lie outside the picture.

```
In[·]:= Manipulate[Module[{f = #1*#2*
        (1 - #1)*(2 - #2) + 2 & }, Show[
        Plot3D[f[x, y], {x, -10, 10}, {y, -10, 10},
        RegionFunction -> If[ww, Function[
        {x, y, z}, x^2 + y^2 <= d^2],
        Function[{x, y, z}, True]], ColorFunction ->
        Function[{x, y, z}, Directive[Opacity[
        0.8], Blue]], Mesh -> False],
        ParametricPlot3D[d*{Cos[u], Sin[u], 0},
        {u, 0, 2*Pi}, ColorFunction ->
        Function[{x, y, z}, Green]], If[tt,
        Graphics3D[{Yellow, PointSize[0.02],
        Point[{x, y, f[x, y]} /. NSolve[
        D[f[x, y], {{x, y}}] == 0 &&
        x^2 + y^2 <= d^2, {x, y}]]}],
        Graphics3D[{}]], If[mm, Graphics3D[
        Line[{{d*Cos[t], d*Sin[t], 0},
        {d*Cos[t], d*Sin[t], f[d*Cos[t],
        d*Sin[t]]}}]], Graphics3D[{}]],
        If[ss, Graphics3D[{Red, PointSize[0.02],
        Point[{x, y, 0} /. NSolve[D[f[x, y],
        {{x, y}}] == 0 && x^2 + y^2 <= d^2,
        {x, y}]]}], Graphics3D[{}]], AxesLabel ->
        {"x", "y", "z"}, Ticks -> None,
        PlotRange -> {{-5, 5}, {-5, 5},
        {-5, 5}}]], {{d, 1, "radius"}, 0.1, 4,
        Appearance -> "Labeled"}, {{ww, True,
        "restrict"}, {True, False}}, {{t, 0, "t"},
        0, 2*Pi}, {{tt, False, "show critical
        values "}, {True, False}}, {{ss,
        True, "show critical points"},
         {True, False}}, {{mm, False, "points
        on boundary "}, {True, False}},
        SynchronousUpdating -> False]
```

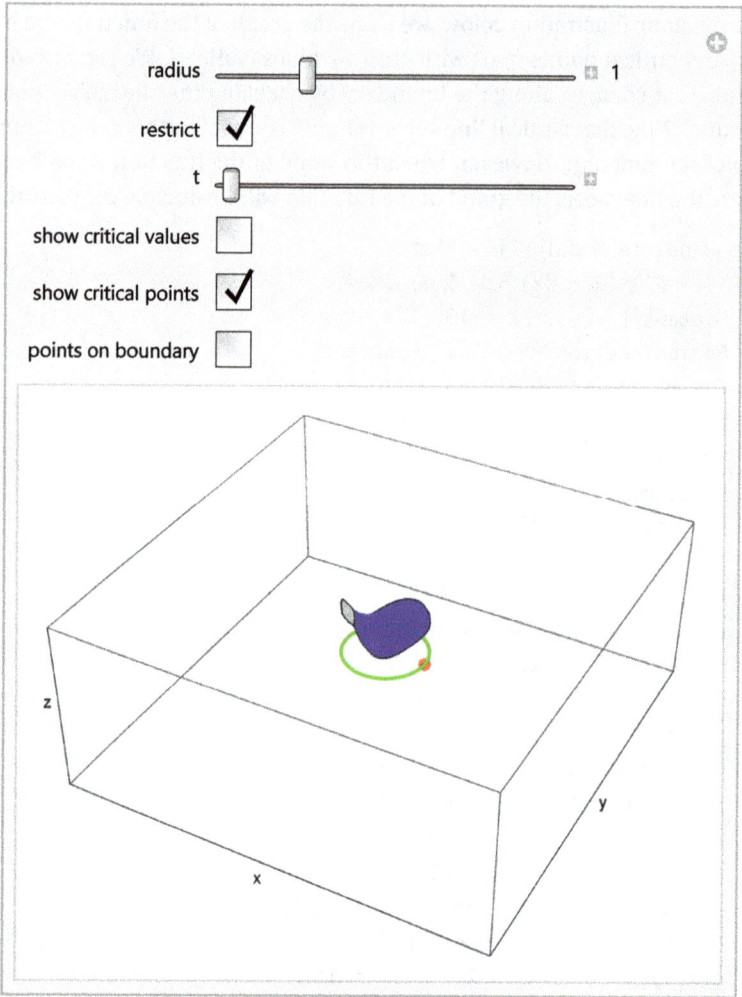

Out[ ]=

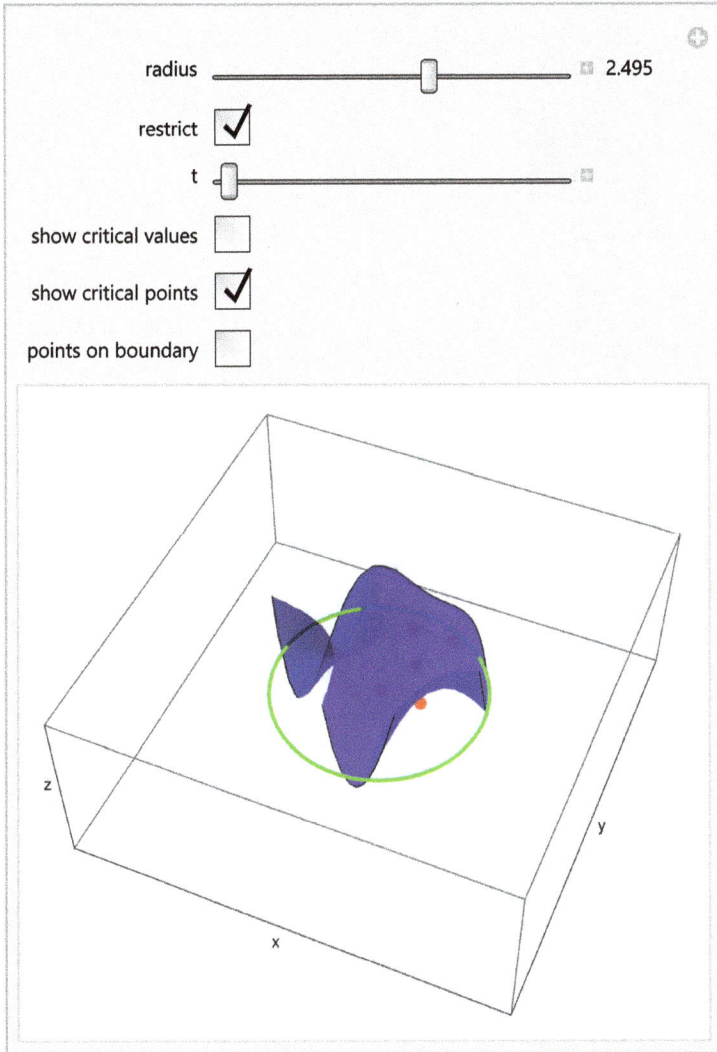

| radius | ──────●────── | ▢ 2.495 |

restrict ☑

t ●──────────── ▢

show critical values ☐

show critical points ☑

points on boundary ☐

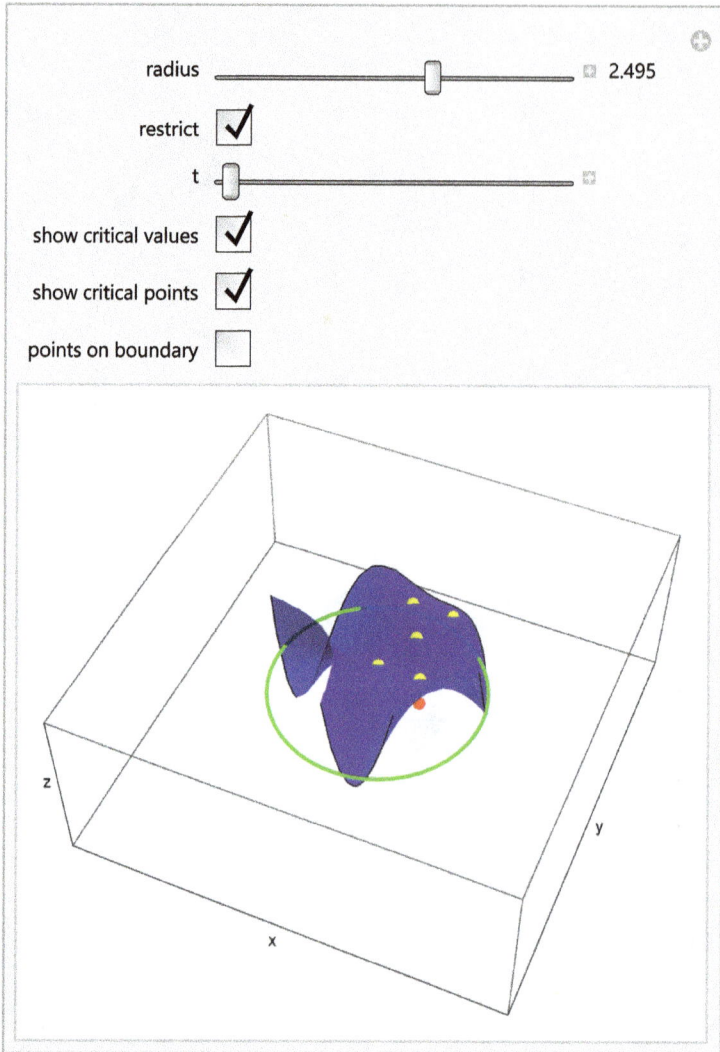

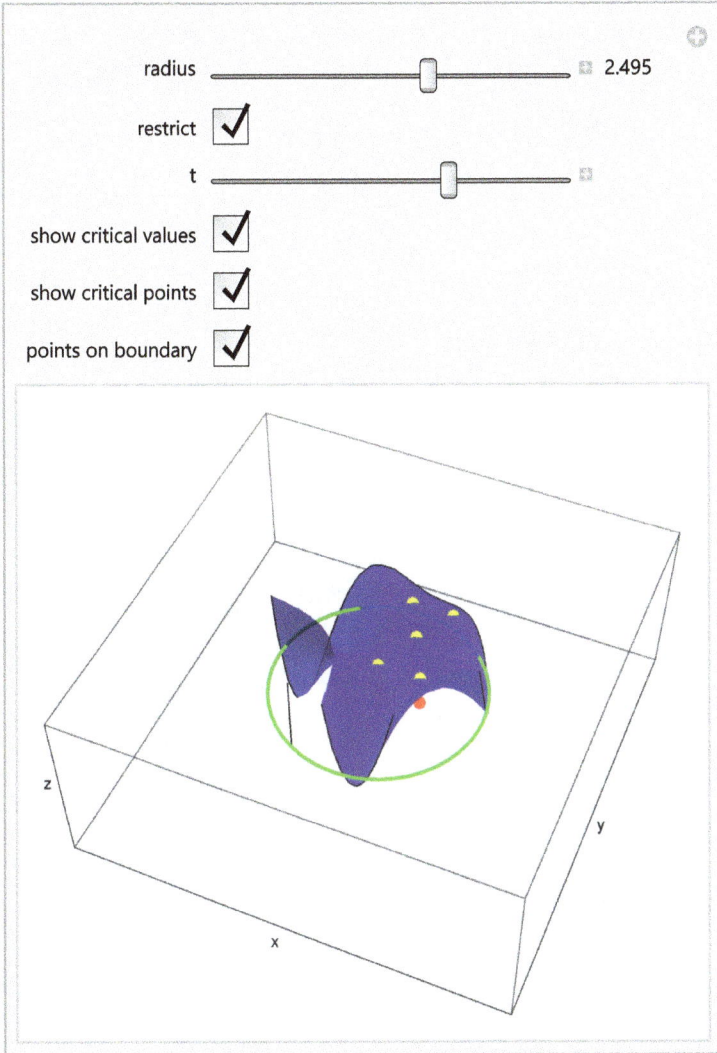

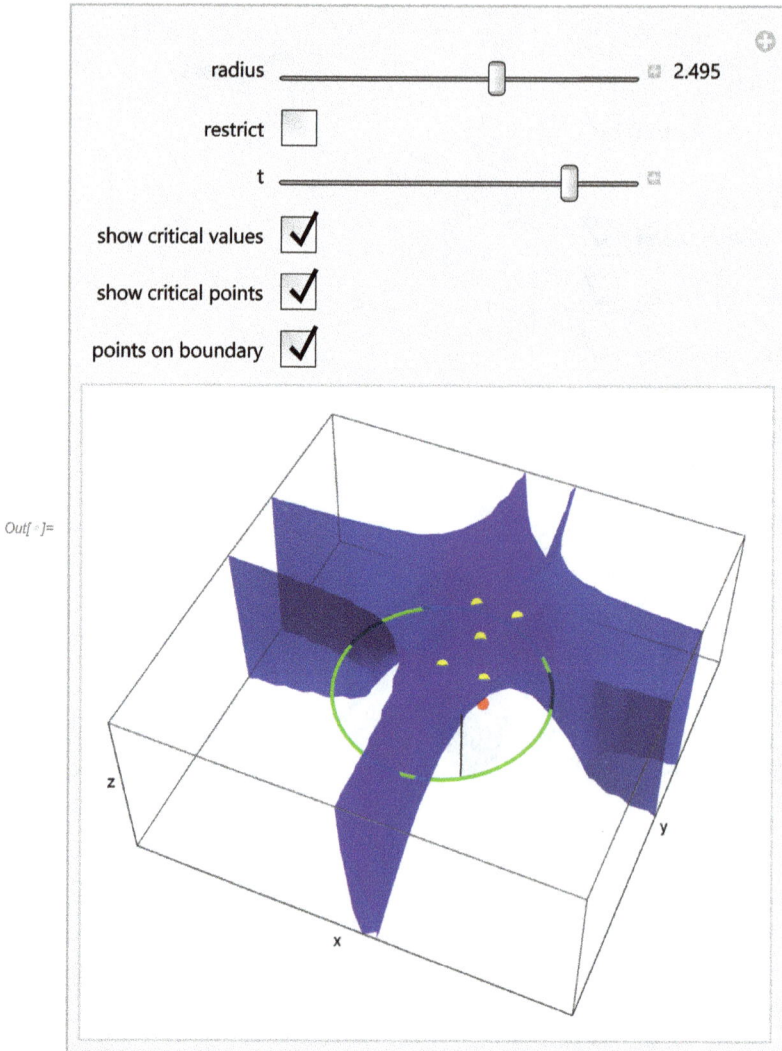

Figure 2.11

In the above case we can also easily solve the problem of determining the extrema of the function on $\mathbb{R}^2$. For $y = x$ we have

```
In[·]:= Expand[f[x, x]]
Out[·]:= 2 + 2 x² - 3 x³ + x⁴
```

This shows that the values of f tend to infinity as we increase x, hence the function is unbounded from above. On the other hand,

In[·]:= Expand[f[x, 1]]
Out[·]:= $2 + x - x^2$

Hence, by increasing x while keeping $y = 1$, we see that the values of f are also un-bounded from below.

The above method can find, under certain conditions, global extrema but in gen-eral it is insufficient to determine the nature of the critical points, that is, which of them are local minima, local maxima, or saddle points. Doing this requires the use of higher order derivatives, which will be considered in the next chapter. Note that even if we could find a disk containing only one critical point, and used the above method to determine the global extrema of the restriction of our function to the disk, it may still not be sufficient to determine the nature of the critical point. The reason for that is the failure of the so-called "the only critical point in town" principle for functions of two or more variables. In the case of differentiable functions of a single variable, the principle states that if a function has only one critical point at which it has a local maximum (minimum), it must have at this point its global maximum (minimum). The proof is easy. In the case of a function of several variables, a point could be a local minimum and the only critical point without being a global minimum. For example, consider the function

In[·]:= Clear[f]; f[x_, y_] := x^2*(1 + y)^3 + y^2

This has only one critical point:

In[·]:= Solve[D[f[x, y], {{x, y}}] == 0, {x, y}]
Out[·]:= {{x -> 0, y -> 0}}

Since the function takes the value 0 at the critical point $(0,0)$ and is nonnegative in a neighborhood of the origin (see the graph below), it has a local minimum there. However, this is not a global minimum because, for example,

In[·]:= f[1, -10]
Out[·]:= -629

In[·]:= Manipulate[Plot3D[{0, f[x, y]}, {x, -a, a},
        {y, -a, a}], {a, 1, 3}, SaveDefinitions -> True]

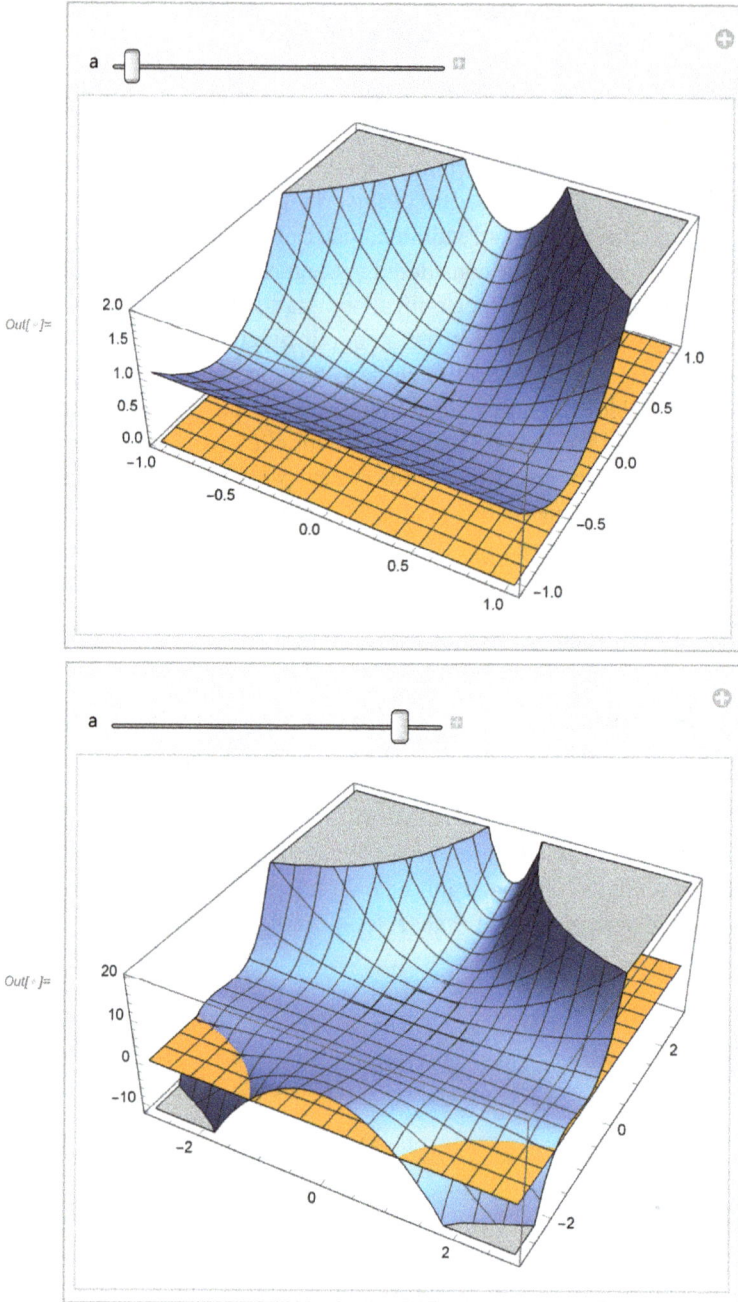

Figure 2.12

2.9 Lagrange multipliers

A situation that occurs in many application is as follows: we have a class C^1 function f : $U \subset \mathbb{R}^n \to \mathbb{R}$, where U is an open set, and we want to find its extrema, not on the whole open set U but only on a subset, where some relation holds between the variables. For example, suppose we are given a differentiable mapping with the same domain, i. e., $g : U \subset \mathbb{R}^n \to \mathbb{R}^m$, and suppose we want to find the extrema of f restricted to the level set of g for some $c \in \mathbb{R}^m$, which is in the range of g (i. e., there exists $x_0 \in U$ such that $g(x_0) = c$). This kind of problem is called a *"constrained optimization problem"*. The function f, whose extrema we wish to find, is often called the *objective function* of the problem and the function g the *constraint*.

Let us consider a simple example. We take $U = \mathbb{R}^n$, and the functions f and g are given by

```
In[·]:= Clear[f, g]; f[x_, y_] := x^2 + y^2
In[·]:= g[x_, y_] := (x - 3)^2 + (y - 1)^2 - 1
```

We want to find the extrema of f on the subset $g^{-1}(0)$. Of course, we could easily solve this problem by parametrizing the constraint (which is a circle) and reducing the problem to one involving only one variable. In fact, this is the method we used to make an illustration below.

Using the natural parametrization of the circle our objective function takes the following form:

```
In[·]:= Simplify[x^2 + y^2 /. {x -> Cos[t] + 3,
          y -> Sin[t] + 1}]
Out[·]:= 11 + 6 Cos[t] + 2 Sin[t]

In[·]:= Show[Plot3D[x^2 + y^2, {x, 1, 5},
          {y, -1, 3}, ColorFunction -> Function[
          {x, y}, Directive[Opacity[0.2], Blue]],
          Mesh -> False], ParametricPlot3D[
          {Cos[t] + 3, Sin[t] + 1, 2*Sin[t] +
          6*Cos[t] + 11}, {t, 0, 2*Pi},
          ColorFunction -> Function[{x, y}, Red]],
          ParametricPlot3D[{Cos[t] + 3, Sin[t] + 1,
          0}, {t, 0, 2*Pi}, ColorFunction ->
          Function[{x, y}, Green]]]
```

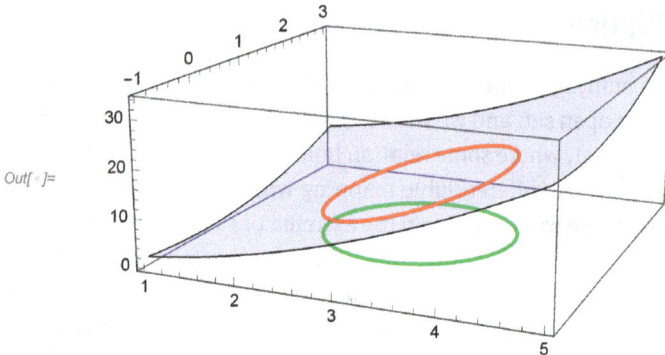

Out[ ]=

Figure 2.13

The level set of g is a circle in the xy-plane given by the equation $g = 0$ (green). Our function f can be viewed as a function defined on this circle, whose graph is another circle (red) consisting of the part of the graph of the function f which lies above the first circle. Looking at the graph, we see that there is a single minimum on the left and a single maximum on the right.

There is another way to approach this problem. We illustrate it below.

```
In[·]:= Manipulate[Module[{f = #1^2 + #2^2 & ,
         g = (#1 - 3)^2 + (#2 - 1)^2 - 1 & , pts},
         pts = {x, y} /. Solve[{g[x, y] == 0,
         D[f[x, y], {{x, y}}] == l*D[g[x, y],
         {{x, y}}]}, {x, y, 1}]; Show[ContourPlot[
         f[x, y] == a^2, {x, -5, 5}, {y, -5, 5},
         ColorFunction -> Function[{x, y},
         Magenta]], Graphics[If[u, {Magenta,
         Arrow[{{a*Cos[t], a*Sin[t]},
         {a*Cos[t], a*Sin[t]} + D[f[x, y],
         {{x, y}}] /. {x -> a*Cos[t], y ->
         a*Sin[t]}}]}, {}]], ContourPlot[g[x, y]
         == 0, {x, -5, 5}, {y, -5, 5},
         ColorFunction -> Function[{x, y},
         Green]], Graphics[If[u, {Green, Arrow[
         {{Cos[s] + 3, Sin[s] + 1}, {Cos[s] +
         3, Sin[s] + 1} + D[g[x, y], {{x, y}}]
         /. {x -> Cos[s] + 3, y -> Sin[s] + 1}}]},
         {}]], Graphics[If[v, {Blue, PointSize[0.02],
         Point[pts]}, {}]], PlotRange -> 6]],
         {{a, 0.1, "a"}, 0, 5}, {{t, 0,
```

```
"gradient of f"}, 0, 2*Pi}, {{s, 0,
"gradient of g"}, 0, 2*Pi}, {{v,
False, "show extrema"}, {True, False}},
{{u, False, "show arrows"}, {True, False}},
SaveDefinitions -> True]
```

Out[ ]=

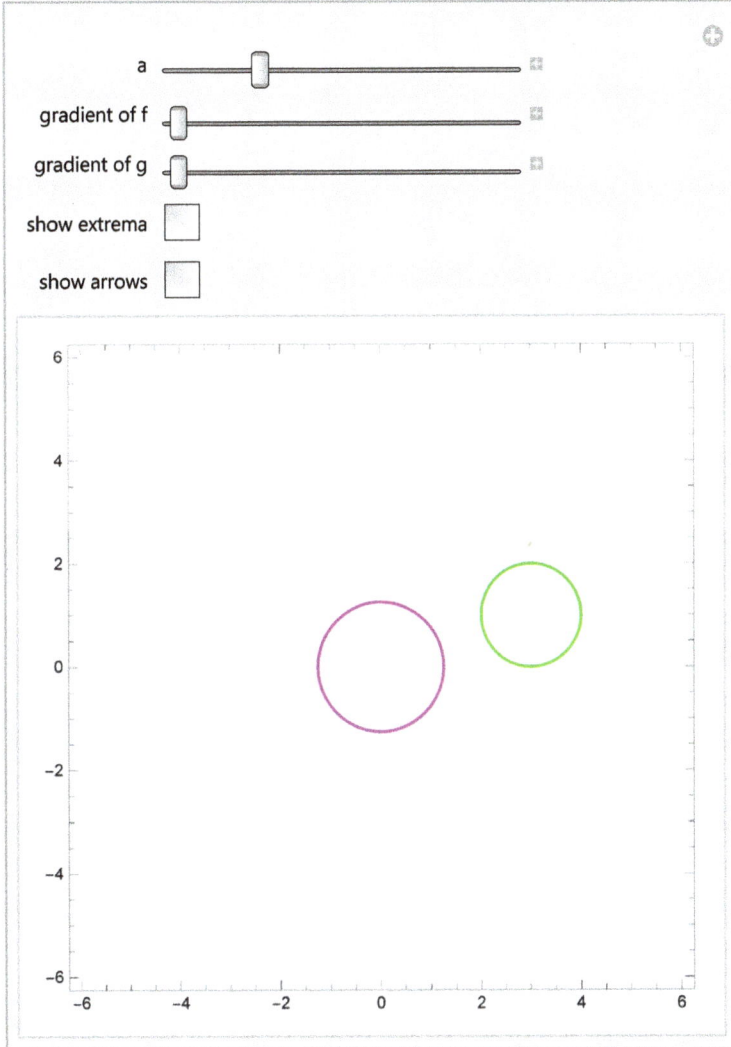

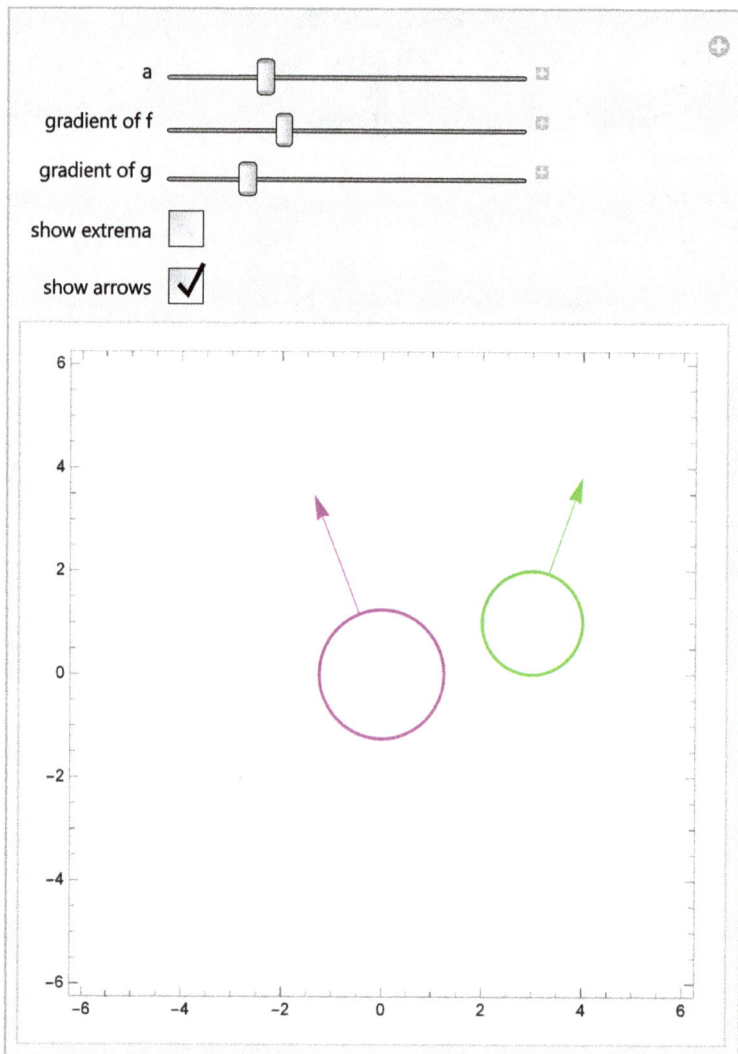

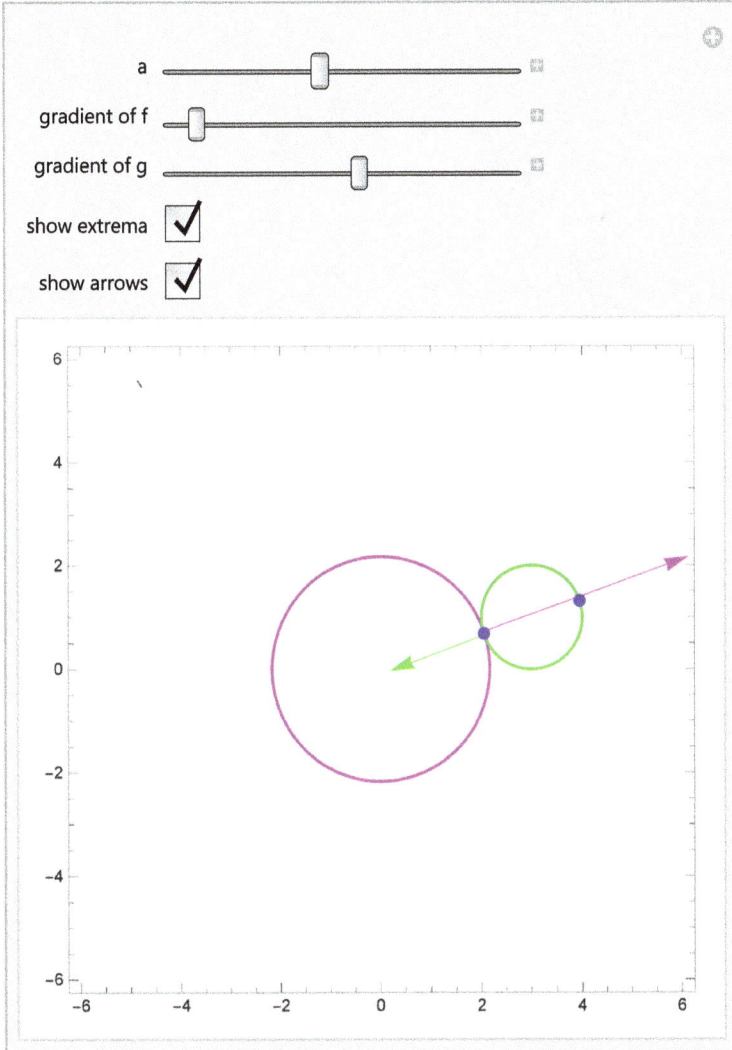

Out[ ]=

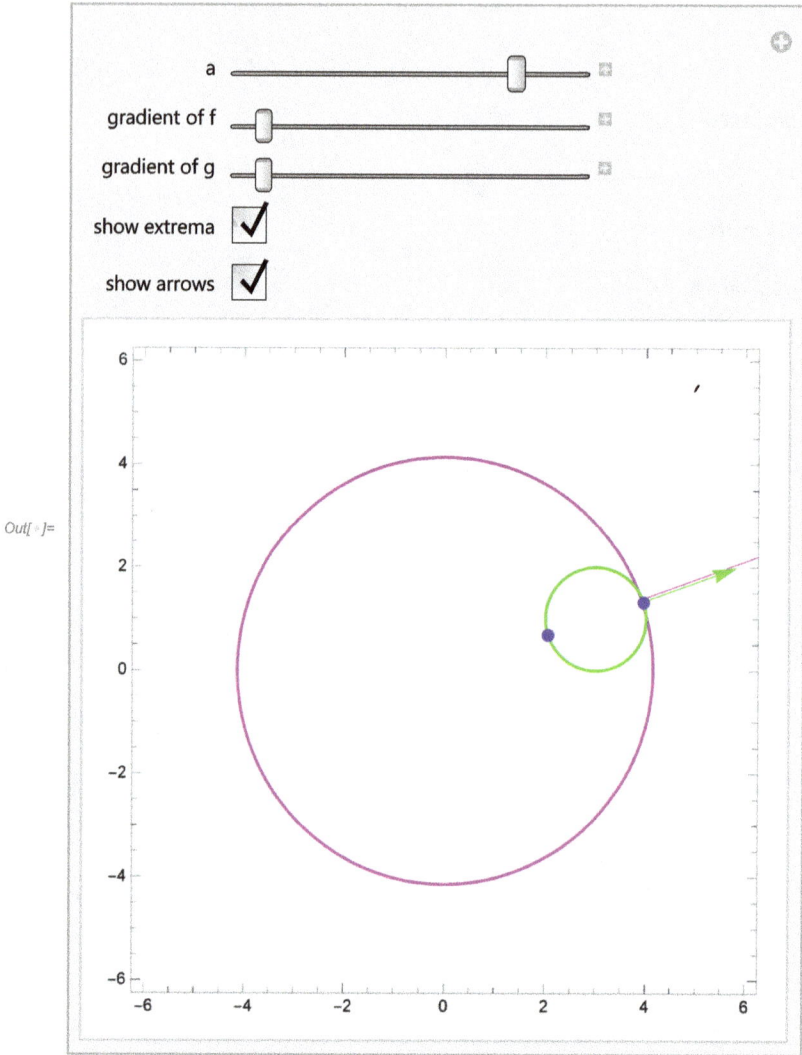

Figure 2.14

Here the green circle is the level curve $g(x, y) = 0$ on which we try to find the extrema of the function f. The pink circle represents the level curve $f(x, y) = a^2$, and we vary the parameter a. Clearly, the union of such circles over all $a \geq 0$ is the entire space $\mathbb{R}^2$. For small values of a, the pink circle is disjoint from the green one, which means that there are no points on the green circle (level curve of g) at which f is equal to a^2. However, as we increase the radius of the pink circle a, the pink circle will first meet the green one at one point, then will intersect it in two, then touch it again at one point and again become disjoint from it. It is now clear that the point on the green circle, where the pink circle touches it for the first time, is where the function f has its minimum, the second point of tangency is where f attains its maximum. Clearly, the two circles have to be tangent at these points, that is, their tangent vectors at these points must be scalar multiples of one another. But since we know that the gradient of a function is always perpendicular to any level curve of the function, the gradients of f and g at these points must be scalar multiples of one another. This allows us to find both points. We simply need to solve three simultaneous equations in three unknowns: one being the constraint equation $g(x, y) = 0$ and the other two given by $\nabla f(x, y) = \lambda \nabla g(x, y)$, where λ is a parameter known as the *Lagrange multiplier*. Since we actually do not normally need to know its value, we can eliminate it:

```
In[·]:= pts = N[{x, y} /. Solve[{g[x, y] == 0,
        D[f[x, y], {{x, y}}] == l*D[g[x, y],
        {{x, y}}]}], {x, y, 1}]]
Out[·]:= {{2.05132, 0.683772}, {3.94868, 1.31623}}
```

This gives us the two points. To find out which is the maximum and which the minimum, we simply compute the value of f at both:

```
In[·]:= N[Apply[f, pts, {1}]]
Out[·]:= {4.67544, 17.3246}
```

which shows that the first point is the minimum and the second is the maximum.

When will the method not work? Clearly, the crucial condition is that the level curve should have a nonzero tangent vector at the extreme points. For example, consider the same function f as above but now let $g(x, y) = (y - 1)^3 - x^2$. Clearly (see the dynamic illustration below), the minimum of f is at attained at the point $(0, 1)$ but the method of Lagrange multipliers finds that there are no solutions:

```
In[·]:= f[x_, y_] := x^2 + y^2
In[·]:= g[x_, y_] := (y - 1)^3 - x^2

In[·]:= Solve[{g[x, y] == 0, D[f[x, y],
        {{x, y}}] == l*D[g[x, y], {{x, y}}]},
        {x, y, 1}, Reals]
Out[·]:= {}

In[·]:= Manipulate[Show[ParametricPlot[
        {t^3, t^2 + 1}, {t, -1, 1}],
        ParametricPlot[{a*Cos[u], a*Sin[u]},
        {u, 0, 2*Pi}], PlotRange ->
        {-1.5, 1.5}, AxesOrigin ->
        {{0, 0}, {-1.5, 2.5}}], {a, 0.5, 1.5}]
```

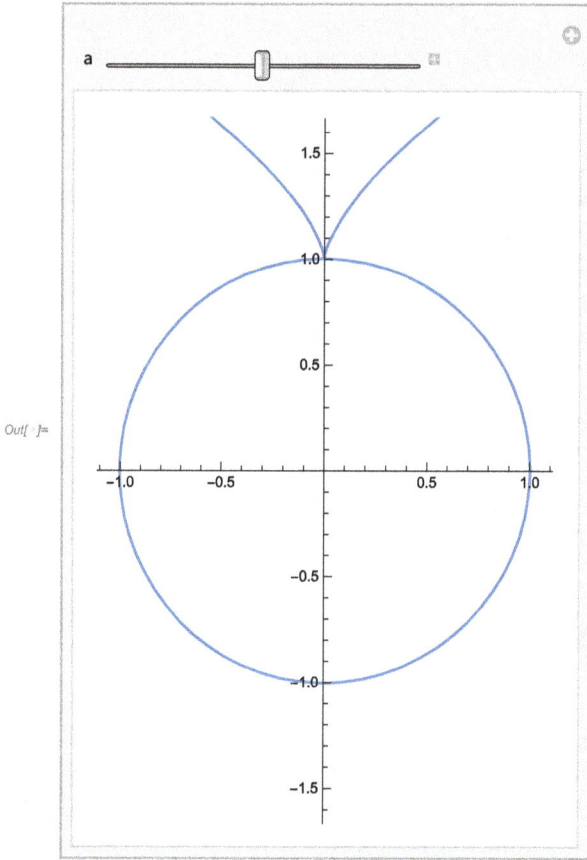

Figure 2.15

Essentially the same approach can be used in situations where we have several constraints, which can be considered as a level set of a mapping $g : U \rightarrow \mathbb{R}^m$. We can state this as a theorem (see [10, Section 3.4]).

Theorem 6. *Let* $f \in C^1(U, \mathbb{R})$ *and let* $g = (g_1, g_2, \ldots, g_m) \in C^1(U, \mathbb{R}^m)$, *where* U *is an open set in* $\mathbb{R}^n$, $n > m$. *Let* $M = \{z \in U, g(z) = 0\}$. *Let* $p \in M$ *and suppose that the linear transformation* $Dg(p)$ *is an epimorphism. Then if* f *has an extremum at a point* $p \in M$, *then there exist numbers* $\lambda_1, \lambda_2, \ldots, \lambda_m$ *such that*

$$\nabla f(p) = \sum_{i=1}^{m} \lambda_i \nabla g_i(p).$$

The condition that $Dg(p)$ is an epimorphism is equivalent to the condition that the Jacobian matrix of g at p has maximal rank m. The implicit function theorem (see [14, Section 6.8 and Remark 6.8.4]) implies that the level set $g^{-1}(0)$ is a smooth manifold, and hence has a tangent vector space at each point.

This theorem gives us a procedure for finding constrained extrema, if the epimorphism condition is satisfied, and if we know that an extremum exists (for example, if the set defined by the constraint is compact). In this situation we solve the above set of equations and find all the solutions, which are the candidates for extrema. We then compute the values of f at these points. If we know that a maximum (minimum) exists, the largest (smallest) of the computed values must be that extremum.

2.9.1 Example 8

Find the extrema of the function $f : \mathbb{R}^3 \to \mathbb{R}$, where

In[·]:= f[x_, y_, z_] := x*y*z

subject to the conditions $g_1 = g_2 = 0$ with

In[·]:= g1[x_, y_, z_] := 1 - x - y - z
In[·]:= g2[x_, y_, z_] := 1 - x^2 - y^2 - z^2

We can visualize the set of points satisfying both constraints.

In[·]:= ParametricPlot3D[{{x, y, 1 - x - y},
        {x, y, Sqrt[1 - x^2 - y^2]},
        {x, y, -Sqrt[1 - x^2 - y^2]}},
        {x, -1, 1}, {y, -1, 1}, Mesh -> False]

Out[·]=

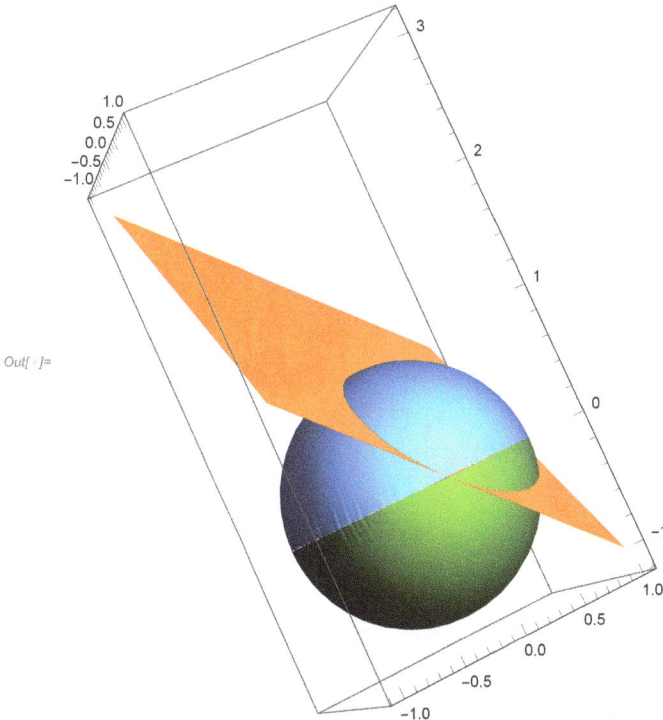

Figure 2.16

From the picture above we see that this set is a circle, which is a smooth manifold, but let us verify that the condition of the theorem is satisfied. We need to check the rank of the matrix:

In[·]:= mm = D[{g1[x, y, z], g2[x, y, z]},
     {{x, y, z}}]
Out[·]:= {{-1, -1, -1}, {-2 x, -2 y, -2 z}}

We cannot rely on **Mathematica**® for this, because its rank function MatrixRank always considers only the generic case, thus:

In[·]:= MatrixRank[mm]
Out[·]:= 2

In fact, the matrix has rank 1 when $x = y = z$ and rank 2 otherwise

In[·]:= MatrixRank[mm /. {y -> x, z -> x}]
Out[·]:= 1

We can prove that this is the only case by using

In[·]:= Reduce[a*mm[[1]] + b*mm[[2]] == 0
     && a^2 + b^2 != 0, {x, y, z}]
Out[·]:= b ≠ 0 && x == $-\dfrac{a}{2b}$ && y == x && z == x && $a^2 + b^2 \neq 0$

Inspecting this we see that either both a and b are 0 or $x = y = z$. However, no point with $x = y = z$ satisfies both constraints

In[·]:= Solve[{g1[x, x, x] == 0, g2[x, x, x] == 0}, x]
Out[·]:= {}

So now we can use the *method of Lagrange multipliers*. We know that the function f is defined on a compact set, hence it will achieve both of its extrema somewhere, and these will have to be at some points found by the method of Lagrange multipliers. Let us apply it:

In[·]:= sols = Solve[{D[f[x, y, z],
     {{x, y, z}}] == 11*D[g1[x, y, z],
     {{x, y, z}}] + 12*D[g2[x, y, z],
     {{x, y, z}}], g1[x, y, z] == 0,
     g2[x, y, z] == 0}, {x, y, z}, {11, 12}]
Out[·]:= $\left\{\left\{x \to -\dfrac{1}{3}, y \to \dfrac{2}{3}, z \to \dfrac{2}{3}\right\}, \{x \to 0, y \to 0, z \to 1\},\right.$
$\{x \to 0, y \to 1, z \to 0\}, \left\{x \to \dfrac{2}{3}, y \to -\dfrac{1}{3}, z \to \dfrac{2}{3}\right\},$
$\left.\left\{x \to \dfrac{2}{3}, y \to \dfrac{2}{3}, z \to -\dfrac{1}{3}\right\}, \{x \to 1, y \to 0, z \to 0\}\right\}$

Note that we used a version of Solve which told **Mathematica**® to eliminate λ_1 and λ_2, which we do not need (see section "Eliminating variables" in the **Mathematica**® tutorial). Of course, it was not necessary to do this.

In[·]:= f[x, y, z] /. sols

Out[·]:= $\left\{-\dfrac{4}{27}, 0, 0, -\dfrac{4}{27}, -\dfrac{4}{27}, 0\right\}$

We see that the largest value achieved is 0 (at three points) and the smallest $-4/27$, also at three points.

It could also happen that the level set of constraints contains singular points, where the rank of the Jacobian matrix of the constraints is not maximal. If there is only a finite number of such points, we can deal with them too, simply by computing the value of f at these points and comparing them with the others returned by the method of Lagrange multipliers.

In Mathematica® we do not need to use the above approach explicitly because we can solve the problem by using Maximize and Minimize. For example,

In[·]:= Maximize[{f[x, y, z], g1[x, y, z] ==
  0, g2[x, y, z] == 0}, {x, y, z}]

Out[·]:= {0, {x → 1, y → 0, z → 0}}

In[·]:= Minimize[{f[x, y, z], g1[x, y, z] ==
  0, g2[x, y, z] == 0}, {x, y, z}]

Out[·]:= $\left\{-\dfrac{4}{27}, \left\{x → \dfrac{2}{3}, y → -\dfrac{1}{3}, z → \dfrac{2}{3}\right\}\right\}$

3 Higher order derivatives and the Taylor series

In this chapter we define higher order derivatives of functions of several variables. Then we consider the Taylor series, the way to obtain it in Mathematica®, and apply it to the problem of determining the nature of critical points.

3.1 Higher order partial derivatives

Let $f : U \subset \mathbb{R}^n \to \mathbb{R}$ be a function for which a partial derivative $\partial f / \partial x_i : U \to \mathbb{R}$ is defined. If this function has a partial derivative with respect to x_j at a point $a \in U$, then we define

$$\frac{\partial^2 f}{\partial x_j \partial x_i}(a) = \frac{\partial}{\partial x_j} \frac{\partial f}{\partial x_i}(a).$$

Continuing in this way, we can inductively define all *higher order partial derivatives*.

As before, in Mathematica® we have two notions of partial derivative, that of an expression and that of a function (see also [5, Section 5.2]). For example, here is the third order partial derivative with respect to x, y, z of the expression $x^y z + xyz$:

> *In[·]:=* D[x^y*z + x*y*z, x, y, z]
> *Out[·]:=* $1 + x^{-1+y} + x^{-1+y} y \operatorname{Log}[x]$

When differentiating with respect to the same variable several times, we can use the form

> *In[·]:=* D[x^y*z + x*y*z, {x, 3}]
> *Out[·]:=* $x^{-3+y} (-2 + y)(-1 + y) y z$

which is the same as

> *In[·]:=* D[x^y*z + x*y*z, x, x, x]
> *Out[·]:=* $x^{-3+y} (-2 + y)(-1 + y) y z$

An alternative approach is to differentiate a function by means of Derivative, e. g.,

> *In[·]:=* Derivative[0, 0, 1][#1*#2*#3 &]
> *Out[·]:=* #1 #2 &

Note that Mathematica® automatically assumes that the "mixed" derivatives of an undefined function are equal:

> *In[·]:=* Clear[f]; D[f[x, y], x, y]
> *Out[·]:=* $f^{(1,1)}[x, y]$

> *In[·]:=* D[f[x, y], y, x]
> *Out[·]:=* $f^{(1,1)}[x, y]$

https://doi.org/10.1515/9783110660395-003

However, this is only true under the assumption that the function is of class C^2, that is, all of its partial derivatives of order two exist at all points of U and are continuous there. Without that condition, the mixed derivatives may not be equal. Let us consider the following example:

```
In[·]:= f[x_, y_] := Piecewise[{{(x*y*(x^2 - y^2))/
         (x^2 + y^2), {x, y} != {0, 0}}}]
```

```
In[·]:= Plot3D[f[x, y], {x, -1, 1},
         {y, -1, 1}, Exclusions -> {{0, 0}}]
```

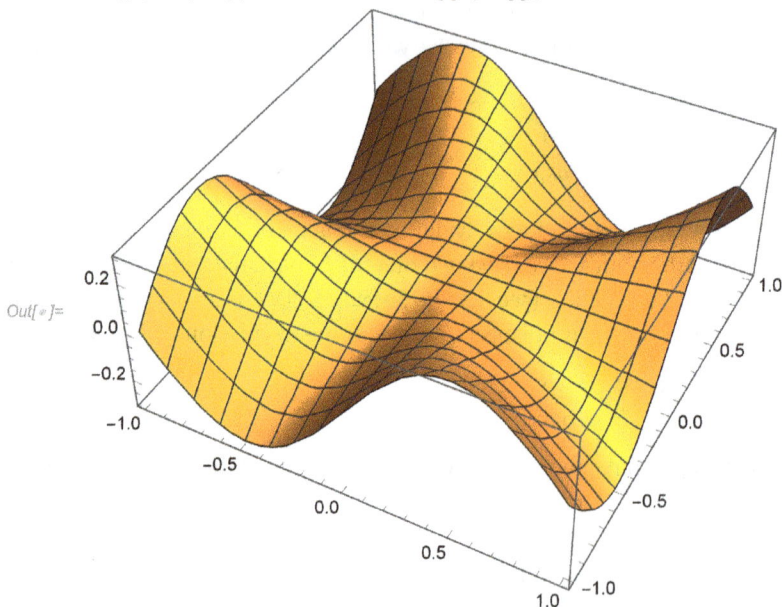

Figure 3.1

We will show that this function has second order partial derivatives, but the mixed derivatives are not equal.

Let us compute the two partial derivatives of f:

```
In[·]:= g[x_, y_] := Simplify[Derivative[1, 0][f][x, y]]
```

```
In[·]:= h[x_, y_] := Simplify[Derivative[0, 1][f][x, y]]
```

At the origin the partial derivatives are clearly zero, and we can show that they are continuous at zero, for instance,

```
In[·]:= Plot3D[g[x, y], {x, -1, 1}, {y, -1, 1},
         Exclusions -> {{0, 0}}]
```

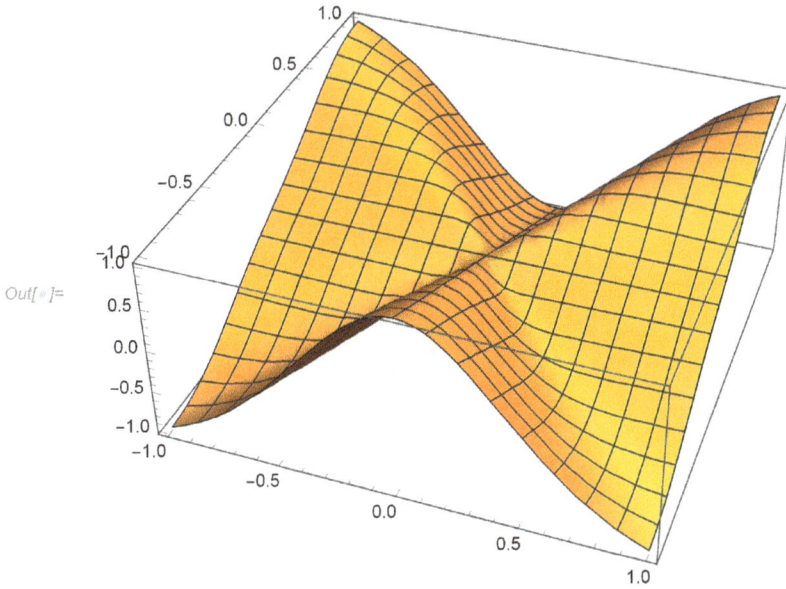

Figure 3.2

Now let us consider the partial derivatives of *g* and *h*:

```
In[·]:= m[x_, y_] := Derivative[0, 1][g][x, y]

In[·]:= n[x_, y_] := Derivative[1, 0][h][x, y]

In[·]:= Plot3D[m[x, y], {x, -1, 1}, {y, -1, 1},
       Exclusions -> {{0, 0}}]
```

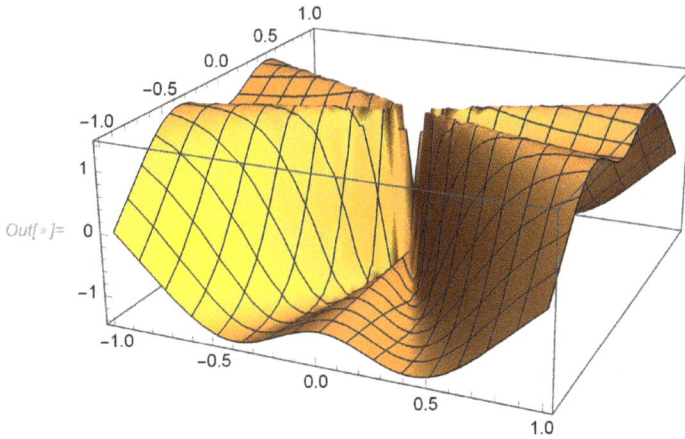

Figure 3.3

```
In[·]:= Plot3D[n[x, y], {x, -1, 1}, {y, -1, 1},
          Exclusions -> {{0, 0}}]
```

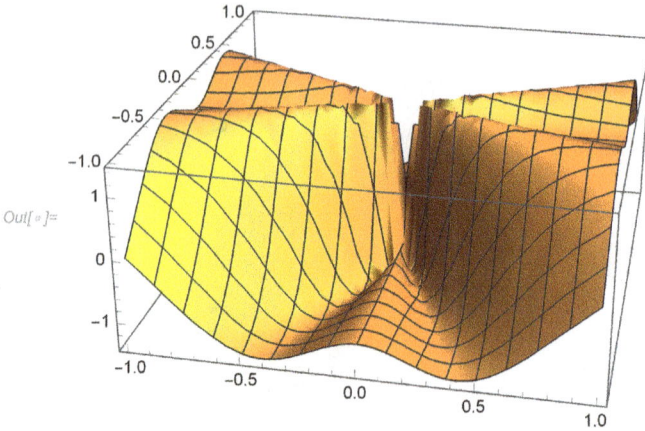

Out[ ]=

Figure 3.4

This time Mathematica®'s computations of partial derivatives are correct only outside $(0, 0)$. We can clearly see that the second order partial derivatives are not continuous at $(0, 0)$. We will show that they are also not equal to 0 at the origin and they are not equal to each other. We need to compute them without using Mathematica®'s second order derivatives. We first compute $\partial g/\partial y(0, 0)$ and then $\partial h/\partial x(0, 0)$:

```
In[·]:= Limit[(g[0, k] - g[0, 0])/k, k -> 0]
Out[·]:= -1
```

```
In[·]:= Limit[(h[k, 0] - h[0, 0])/k, k -> 0]
Out[·]:= 1
```

Thus we see that in this case the mixed second order partial derivatives at the origin are not equal.

Let us now consider the meaning of higher order derivatives. Let $f : U \subset \mathbb{R}^n \to \mathbb{R}^m$ be differentiable on U. We know that the derivative at a point is a mapping $U \to L(\mathbb{R}^n, \mathbb{R}^m)$, where $L(\mathbb{R}^n, \mathbb{R}^m)$ denotes the set of all linear maps $\mathbb{R}^n \to \mathbb{R}^m$, which can be identified with the set $M_{n \times m}(\mathbb{R})$ of real matrices. This can be identified with the space $\mathbb{R}^{nm}$ and hence the derivative is a map $U \to \mathbb{R}^{nm}$. If this is differentiable, its derivative at a point is a linear map in $L(\mathbb{R}^n, L(\mathbb{R}^n, \mathbb{R}^m)) = L(\mathbb{R}^n, \mathbb{R}^{nm})$. There is a natural identification of $L(\mathbb{R}^n, L(\mathbb{R}^n, \mathbb{R}^m))$ with $BL(\mathbb{R}^n \times \mathbb{R}^n, \mathbb{R}^m)$, the set of bilinear maps $\mathbb{R}^n \times \mathbb{R}^n \to \mathbb{R}^m$, which takes a mapping T to $(u, v) \to T(u)(v)$. In an analogous way, we can define the third order derivative at a point as a trilinear mapping, etc.

To see more clearly how this works in Mathematica®, consider a function of two variables $f : \mathbb{R}^2 \to \mathbb{R}$. The first order derivative can be identified with the gradient

In[·]:= `Clear[f,v]; D[f[x, y], {{x, y}}]`
Out[·]:= $\left\{f^{(1,0)}[x, y], f^{(0,1)}[x, y]\right\}$

This linear mapping is given by the dot product:

In[·]:= `D[f[x, y], {{x, y}}] . {u, v}`
Out[·]:= $v\,f^{(0,1)}[x, y] + u\,f^{(1,0)}[x, y]$

The second order derivative can be identified with the matrix (called the *Hessian*)

In[·]:= `D[f[x, y], {{x, y}, 2}]`
Out[·]:= $\left\{\left\{f^{(2,0)}[x, y], f^{(1,1)}[x, y]\right\}, \left\{f^{(1,1)}[x, y], f^{(0,2)}[x, y]\right\}\right\}$

This acts on pairs of vectors by

In[·]:= `{u, v} . D[f[x, y], {{x, y}, 2}] . {s, t}`
Out[·]:= $t\,(v\,f^{(0,2)}[x, y] + u\,f^{(1,1)}[x, y]) +$
$\qquad s\,(v\,f^{(1,1)}[x, y] + u\,f^{(2,0)}[x, y])$

Note that the Hessian matrix returned by Mathematica® is symmetric. (We will explain this below.) As is well known from linear algebra, to every symmetric bilinear function there corresponds a unique quadratic form given by

In[·]:= `Expand[{u, v} . D[f[x, y], {{x, y}, 2}] . {u, v}]`
Out[·]:= $v^2\,f^{(0,2)}[x, y] + 2\,u\,v\,f^{(1,1)}[x, y] + u^2 f^{(2,0)}[x, y]$

In the case of the third order derivative, the Hessian matrix is replaced by the tensor

In[·]:= `D[f[x, y], {{x, y}, 3}]`
Out[·]:= $\Big\{\Big\{\big\{f^{(3,0)}[x, y], f^{(2,1)}[x, y]\big\},$
$\qquad \big\{f^{(2,1)}[x, y], f^{(1,2)}[x, y]\big\}\Big\},$
$\qquad \Big\{\big\{f^{(2,1)}[x, y], f^{(1,2)}[x, y]\big\},$
$\qquad \big\{f^{(1,2)}[x, y], f^{(0,3)}[x, y]\big\}\Big\}\Big\}$

In fact, this represents a symmetric trilinear function which corresponds to a cubic form. Rather than trying to describe the associated cubic form, we now turn to a general formula in which all these forms appear. This formula is the multi-dimensional Taylor formula [10, p. 160].

Let $f : U \subset \mathbb{R}^n \to \mathbb{R}$ be a k-times differentiable function on an open set U. Let $a \in U$ and let $r > 0$ be such that $B(a, r)$ (the ball of radius r centered at a) is contained in U. Let $D^i f(a)$ denote the ith derivative of f at a, which we view as a i-linear map $\mathbb{R}^n \times \cdots \times \mathbb{R}^n \to \mathbb{R}$. We will write $Df(a)h^k$ to mean $Df(a)(h, \ldots, h)$. The multi-variable *Taylor theorem* says that for every vector h with $\|h\| < r$ we have

$$f(a + h) = f(a) + Df(a)h + \frac{1}{2!}D^2f(a)h^2 + \cdots + \frac{1}{k!}D^kf(a)h^k + R(h),$$

where $R(h)/\|h\|^k \to 0$ as $h \to 0$.

Let us now see how we can quickly obtain a multi-variable Taylor expansion with Mathematica®. We fix a and h and consider the function of one variable t, $t \to f(a+ht)$.

Here t is in some open interval containing $[0, 1]$. We can then expand the function as a Taylor series in t. For example, let $a = (x, y)$ and $h = (u, v)$. Then

In[·]:= Clear[f]; f[x + t*u, y + t*v] + O[t]^3
Out[·]:= $f[x, y] + (v\, f^{(0,1)}[x, y] + u\, f^{(1,0)}[x, y])\, t +$
$$\frac{1}{2}\,(v^2\, f^{(0,2)}[x, y] + 2\, u\, v\, f^{(1,1)}[x, y] + u^2 f^{(2,0)}[x, y])\, t^2 + O[t]^3$$

On the other hand, using Taylor's theorem above, we get

$$f(x + tu, y + tv) = f(x, y) + t\, Df(x, y)(u, v) + \frac{1}{2!}\, t^2\, D^2 f(x, y)(u, v)^2 + R(t(u, v)).$$

By the uniqueness of Taylor formula for functions of one variable, we see that we get the coefficient of the Taylor series for several variables by using the Taylor series for functions of one variable. In particular,

In[·]:= SeriesCoefficient[f[x + t*u, y + t*v] + O[t]^3, 2]
Out[·]:= $\frac{1}{2}\,(v^2 f^{(0,2)}[x, y] + 2\, u\, v\, f^{(1,1)}[x, y] + u^2 f^{(2,0)}[x, y])$

gives us the Hessian quadratic form we obtained earlier.

3.1.1 Example 1

Find the fifth Taylor polynomial at $(0, 0, 0)$ of the function

In[·]:= f[x_, y_, z_] := x^2*y*z*Exp[x*y*z]

In[·]:= Normal[f[t*u, t*v, t*w] + O[t]^5] /. t -> 1
Out[·]:= $u^2\, v\, w$

Remark. Mathematica®'s built-in function Series will compute the Taylor series of a function of several variables but it will produce monomials in each variable up to the degree specified for this variable. It makes it more difficult to calculate the Taylor polynomial of a specified total degree. In the above case, the following will work:

In[·]:= Normal[Series[f[u, v, w], {u, 0, 2},
        {v, 0, 2}, {w, 0, 2}]]
Out[·]:= $u^2\, v\, w$

3.2 Local extrema

We know that a necessary condition for a function $f : U \subset \mathbb{R}^n \to \mathbb{R}$ to have a local extremum at a point $a \in U$ is that a is a critical point, that is, $\nabla f(a) = 0$. We will now use the multivariate Taylor theorem to find sufficient conditions, under the assumption that f is of class C^2, i. e., all second order partial derivatives exist and are continuous.

The criterion will involve the Hessian matrix, or, equivalently, symmetric bilinear form (or, equivalently, quadratic form). The *Hessian matrix* of f at a is the $n \times n$ matrix

$$Hf(a) = \left(\frac{\partial^2 f}{\partial x_i \partial x_j}(a) \right)^n_{i,j=1}$$

For example,

In[·]:= D[f[x, y], {{x, y}, 2}]
Out[·]:= {{f$^{(2,0)}$[x, y], f$^{(1,1)}$[x, y]}, {f$^{(1,1)}$[x, y], f$^{(0,2)}$[x, y]}}

Since we are assuming that f is of class C^2, this matrix is symmetric.

Now recall that a real symmetric matrix H is said to be *positive definite* (*negative definite*) if all of its eigenvalues are strictly positive (negative) or, equivalently, if its associated quadratic form $x^{\text{tr}} Hx$ takes always strictly positive (negative) values for all nonzero vectors x. If we replace the requirement of strict positivity (negativity) to be a weak one (i. e., allow some eigenvalues to be 0), then the matrix H is called positive (negative) *semidefinite*. If a matrix is neither positive nor negative semidefinite (at least two nonzero eigenvalues have different signs), then it is called *indefinite*.

The main result of this section is the following theorem (see also [10, p. 173]).

Theorem 7. *If $f : U \subset \mathbb{R}^n \to \mathbb{R}$ is of class C^2, $a \in U$ is a critical point of f, and the Hessian $Hf(a)$ is positive (negative) definite, then a is a (strict) relative minimum (maximum) of f. If $Hf(a)$ is indefinite, then a is neither a local maximum nor a local minimum.*

The proof uses the Taylor theorem, and we omit it except for the last statement. For simplicity, let us assume that $n = 2$, the general argument is completely analogous. Suppose that $Hf(a)$ is indefinite. It must then have two nonzero eigenvalues of opposite signs. Let them be $\lambda_1 > 0$ and $\lambda_2 < 0$. Let u be a nonzero eigenvector of λ_1 and v be a nonzero eigenvector of λ_2. Let $a = (a_1, a_2)$ and $u = (u_1, u_2)$. Then

In[·]:= f[a1 + t*u1, a2 + t*u2] - f[a1, a2] + O[t]^3
Out[·]:= $\left(u2\, f^{(0,1)}[a1, a2] + u1\, f^{(1,0)}[a1, a2] \right) t\, +$
$\quad \frac{1}{2} \left(u2^2\, f^{(0,2)}[a1, a2] + 2\, u1\, u2\, f^{(1,1)}[a1, a2] \right.$
$\quad \left. + u1^2\, f^{(2,0)}[a1, a2] \right) t^2 + O[t]^3$

Since (a_1, a_2) is a critical point, the first term vanishes, and we have

$$f(a_1 + tu_1, a_2 + tu_2) - f(a_1, a_2) = \frac{1}{2} \left(u_1^2 \frac{\partial^2 f}{\partial x^2}(a_1, a_2) \right.$$
$$\left. + 2u_1 u_2 \frac{\partial^2 f}{\partial x \partial y}(a_1, a_2) + u_2^2 \frac{\partial^2 f}{\partial y^2}(a_1, a_2) \right) t^2 + O(t^3).$$

The coefficient of t^2 can be written as

In[·]:= (1/2)*{u1, u2} . (D[f[x, y], {{x, y}, 2}]
     /. {x -> a1, y -> a2}) . {u1, u2}

Out[·]:= $\frac{1}{2}\left(u2\left(u2\, f^{(0,2)}[a1, a2] + u1\, f^{(1,1)}[a1, a2]\right) + u1\left(u2 f^{(1,1)}[a1, a2] + u1\, f^{(2,0)}[a1, a2]\right)\right)$

But since (u_1, u_2) is an eigenvector with eigenvalue λ_1 (which we denote as $l1$ in the code below), the above reduces to

In[·]:= (1/2)*{u1, u2} . {l1*u1, l1*u2}

Out[·]:= $\frac{1}{2}(l1\, u1^2 + l1\, u2^2)$

which is positive. By choosing a sufficiently small t, we now see that $f(a_1 + tu_1, a_2 + tu_2) \geq f(a_1, a_2)$, that is, (a_1, a_2) is not a local maximum. Applying the same argument to the other eigenvalue, we see that it is also not a local minimum, which ends the proof of the statement.

3.2.1 Example 2

Find the local extrema of the function $f : \mathbb{R}^3 \to \mathbb{R}$

In[·]:= f[x_, y_, z_] := -3*x^4 - 2*y^2 -
     z^2 - 2*x*z + y*z

First, we find the critical points

In[·]:= critrules = Solve[D[f[x, y, z],
     {{x, y, z}}] == 0, {x, y, z}, Reals];

In[·]:= crits = {x, y, z} /. critrules

Out[·]:= $\left\{\{0, 0, 0\}, \left\{-\frac{2}{\sqrt{21}}, \frac{4}{7\sqrt{21}}, \frac{16}{7\sqrt{21}}\right\},\right.$
     $\left.\left\{\frac{2}{\sqrt{21}}, -\frac{4}{7\sqrt{21}}, -\frac{16}{7\sqrt{21}}\right\}\right\}$

We next compute the Hessian matrices:

In[·]:= hess = D[f[x, y, z], {{x, y, z}, 2}]
     /. critrules

Out[·]:= $\{\{\{0, 0, -2\}, \{0, -4, 1\}, \{-2, 1, -2\}\},$
     $\left\{\left\{-\frac{48}{7}, 0, -2\right\}, \{0, -4, 1\}, \{-2, 1, -2\}\right\},$
     $\left\{\left\{-\frac{48}{7}, 0, -3\right\}, \{0, -4, 1\}, \{-2, 1, -2\}\right\}\}$

We can compute the eigenvalues of these Hessian matrices by means of the function Eigenvalues, but **Mathematica**® has the function PositiveDefiniteMatrixQ which makes everything even simpler:

```
In[·]:= PositiveDefiniteMatrixQ /@ hess
Out[·]:= {False, False, False}
```

```
In[·]:= NegativeDefiniteMatrixQ /@ hess
Out[·]:= {False, True, True}
```

This tells us that the second and third points are local maxima, but the first case has to be decided separately. Let us have a look at the eigenvalues:

```
In[·]:= N[Eigenvalues[hess[[1]]]]
Out[·]:= {-4.58423, -2.70572, 1.28995}
```

There is one positive and two negative eigenvalues, so the critical point is neither a local minimum nor a local maximum. We could also have checked this by

```
In[·]:= IndefiniteMatrixQ[hess[[1]]]
Out[·]:= True
```

3.2.2 Example 3

Now let us consider an example, where a different approach is needed. Consider the function $g : \mathbb{R}^2 \to \mathbb{R}$ given by

```
In[·]:= g[x_, y_] := (y - x^3)*(y - 2*x^3)
```

There is only one critical point

```
In[·]:= critrules = Solve[D[g[x, y],
           {{x, y}}] == 0, {x, y}, Reals];
```

```
In[·]:= crits = {x, y} /. critrules
Out[·]:= {{0, 0}}
```

However, the Hessian is

```
In[·]:= D[g[x, y], {{x, y}, 2}] /. critrules
Out[·]:= {{{0, 0}, {0, 2}}}
```

```
In[·]:= Eigenvalues[%]
Out[·]:= {2, 0}
```

Thus we have obtained a case which cannot be determined from the Hessian alone. Let us plot the graph of the function g and the two curves $y = x^3$ and $y = 2x^3$ on which the function g takes the value 0.

```
In[·]:= Show[{Plot3D[g[x, y], {x, -1, 1},
           {y, -1, 1}, Mesh -> None,
           ColorFunction -> Function[{x, y, z},
           {Opacity[0.3], Pink}]],
           ParametricPlot3D[{{x, x^3, 0},
           {x, 2*x^3, 0}}, {x, -1, 1}]}]
```

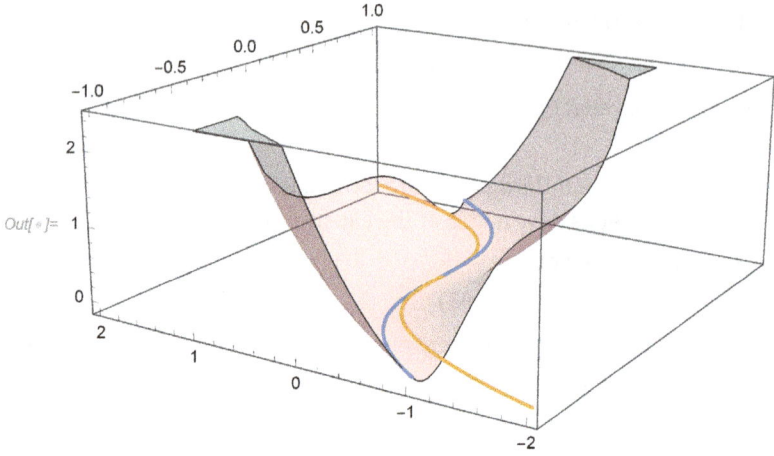

Figure 3.5

```
In[ ]:= Show[ContourPlot[g[x, y], {x, -1, 1},
        {y, -1, 1}, Contours -> 20],
        Plot[{x^3, 2*x^3}, {x, -1, 1},
        PlotStyle -> {Green, Yellow}],
        Graphics[{PointSize[0.02],
        Yellow, Point[{0, 0}]}]]
```

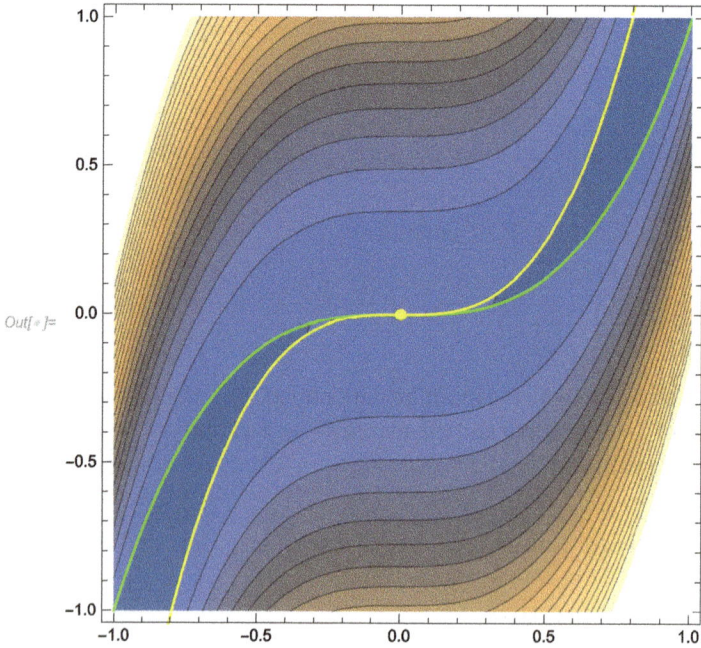

Figure 3.6

We see that two curves separate the plane into four (open) regions, in which the function (by continuity) has to have a constant sign. So we only need to chose one point in each region and check the sign of g at each point. Consider first the region

$$\{(x, y) \in \mathbb{R}^2 \mid x > 0, \, x^3 < y < 2x^3\}.$$

The point $(1, 3/2)$ lies in this region, and we have

In[·]:= Sign[g[1, 3/2]]
Out[·]:= -1

Next, consider the region

$$\{(x, y) \in \mathbb{R}^2 \mid x > 0 \text{ and } y > 2x^3 \text{ or } x < 0 \text{ and } y > x^3\}.$$

The point $(1, 3)$ lies in this region and

In[·]:= Sign[g[1, 3]]
Out[·]:= 1

We have found two regions such that the critical point $(0, 0)$ lies on their boundaries and the function g is negative in one and positive in the other. That, of course, implies that $(0, 0)$ is neither a local minimum nor a local maximum.

4 Elements of topology and Morse theory

In this chapter we will continue the study of functions of many variables. We will apply the methods of the previous chapters to the development of elements of topology and Morse theory.

Since the publication of Milnor's book [11] in 1963, Morse theory has been a standard topic in the education of geometers and topologists. The book itself is still the most popular introductory reference for the subject. Milnor's book begins with an explicit example: the height function on the torus imbedded in the three-dimensional Euclidean space. We will use the same example here. Morse theory is an excellent way to introduce students to more advanced concepts of topology, because, on the one hand, its foundations are very intuitive and closely related to the familiar theory of maxima and minima of a function in $\mathbb{R}^n$ and, on the other hand, because it quickly leads to powerful and sophisticated techniques that bring the student near the frontier of modern research.

Mathematica® is an excellent tool for introducing Morse theory because of its ability to combine sophisticated analytic and algebraic tools with interactive graphics. The latter, of course, are the most useful in the case of surfaces in $\mathbb{R}^3$, but that is precisely the case where the theory is the most complete and hence the most suitable for introducing the subject.

Topology and, in particular, Morse theory have many fascinating applications outside mathematics. We shall not try to describe them but we strongly recommend the excellent book by Edelsbrunner and Harer [4]. It assumes almost no mathematical knowledge, yet it carefully explains everything in this chapter and much more besides.

4.1 Elements of topology

4.1.1 Manifolds

Recall that the most fundamental concept of topology is, naturally, that of a topological space (see Chapter 1, Section 1.1). We have already defined the concept of a continuous map between two topological spaces X and Y (see Chapter 1, Sections 1.1, 1.5). In particular, continuous functions on a topological space are just continuous maps from X to $\mathbb{R}$, where the topology in $\mathbb{R}$ is given by the usual open sets (unions of open intervals).

Once we have the notion of continuous functions on a topological space, it is natural to try to see how much of the usual analysis on an Euclidean space can be done in this more general case. For this purpose, however, general topological spaces have too little structure. To proceed further, we need to restrict ourselves to certain special kinds of topological spaces. Those that most closely correspond to what we intu-

https://doi.org/10.1515/9783110660395-004

itively mean by "space" and which are also the most useful in applications (in physics, etc.) are manifolds, that is, higher-dimensional generalizations of surfaces. Manifolds come in two varieties, without and with a boundary.

A (topological) *manifold M* (without a boundary) is a subset of some $\mathbb{R}^n$ in which for each point $p \in M$ there is an open subset $U_p \subset \mathbb{R}^n$ such that $M \cap U_p$ is homeomorphic to an open subset $\mathbb{R}^d$ (for a fixed $d \leq n$). This gives us local coordinates on M. (It is also useful to define abstract manifolds that are not imbedded in any concrete $\mathbb{R}^n$ by means of an atlas but we shall not need them here). The integer d is called the *dimension of a manifold*. A *differentiable manifold* is a topological manifold where the transition homeomorphisms are actually diffeomorphisms (differentiable homeomorphisms). In the following we shall deal with smooth manifolds.

The *torus* will be our basic example of a manifold. It is a 2-dimensional manifold without a boundary. The following dynamic visualization shows a torus, a point on the torus, and for each point a neighborhood of the point homeomorphic to a disk in $\mathbb{R}^2$. Moving a point in the rectangle causes the image of the point under the standard identification map to move. Note that points lying directly opposite each other on the edges of the rectangle map onto the same point on the torus. Thus the torus can be thought of as a rectangle with opposite edges identified.

```
In[·]:= ttorus[a_, b_, c_][u_, v_] := {(a + b*Cos[v])*Cos[u],
        (a + b*Cos[v])*Sin[u], c*Sin[v]}
```

```
In[·]:= Manipulate[Show[ParametricPlot3D[Evaluate[
        ttorus[2, 1, 1][u, v]], {u, 0, 2*Pi}, {v, 0, 2*Pi},
        ColorFunction -> ({Opacity[0.5], Green} & ),
        Mesh -> False, Ticks -> False, RegionFunction ->
        If[full, Function[{x, y, z}, Norm[{x, y, z} -
        ttorus[2, 1, 1] @@ p] < 1/2], True & ]],
        Graphics3D[{Red, PointSize[0.01],
        Point[ttorus[2, 1, 1] @@ p]}], PlotRange ->
        {{-3, 3}, {-3, 3}, {-3, 3}}], {{p, {0, 0}},
        {-Pi, -Pi}, {Pi, Pi}, Slider2D}, {{full, False,
        "show Euclidean neighborhood "}, {True, False}},
        SaveDefinitions -> True]
```

p

show Euclidean neighbourhood ☐

Out[ ]=

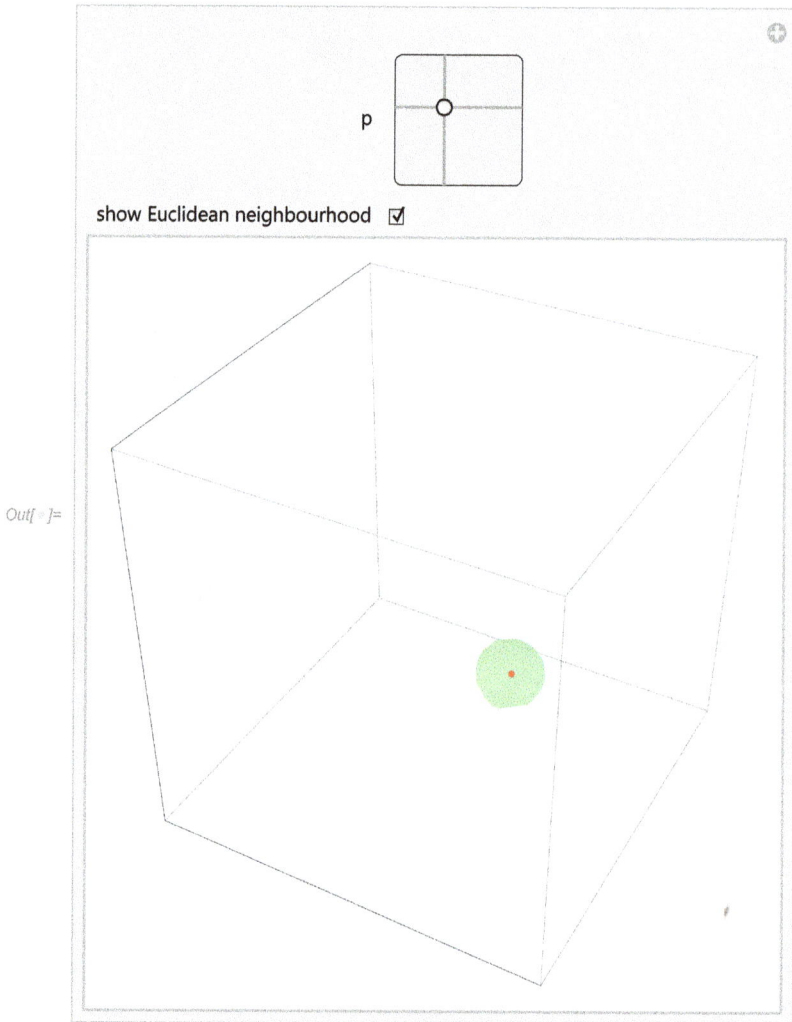

Figure 4.1

The option RegionFunction specifies the region of the graph which satisfies certain condition. In our case, when the box is checked, Mathematica® will draw only points which lie in the neighborhood of the point p on the torus, otherwise it will draw the whole torus.

There are also *manifolds with boundary*, where each point has a neighborhood homeomorphic to the upper-half space $\{(x_1, \ldots, x_d), x_d \geq 0\}$. The disk below is a 2-manifold (surface) with boundary: its boundary is a circle, which is a 1-manifold without boundary. In what follows we shall mainly consider manifolds without a boundary.

```
In[·]:= GraphicsGrid[{{Graphics[{Blue, Disk[{0, 0}, 1]}],
         Graphics[{Blue, Circle[{0, 0}, 1]}]}}]
```

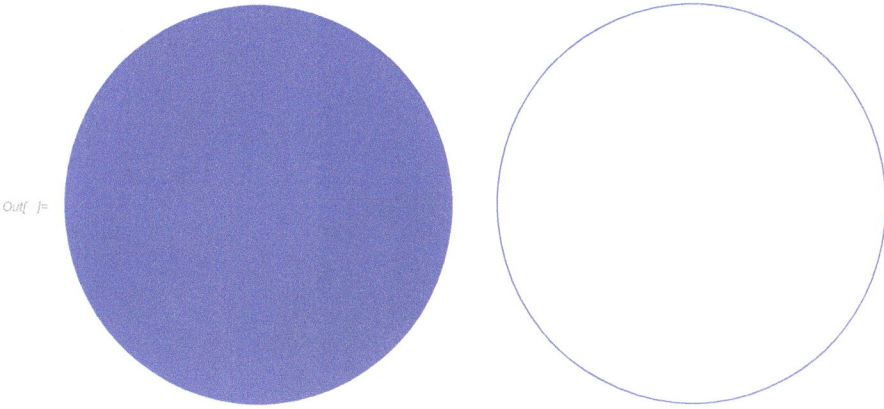

Out[ ]=

Figure 4.2

4.1.2 Classification problem

One of the main (but almost certainly unachievable in full generality) aims of topology
is classification of different kinds of manifolds up to homeomorphism or diffeomorphism. The following dynamic visualization illustrates the notion of homeomorphism.
Changing any of the three parameters a, b, c produces homeomorphic surfaces, each
of which is a torus.

```
In[·]:= Manipulate[ParametricPlot3D[ttorus[a, b, c]
         [u, v], {u, 0, 2*Pi}, {v, 0, 2*Pi}, Mesh -> False,
         Ticks -> False], {{a, 1, "a"}, 0.1, 5},
         {{b, 1, "b"}, 0.1, 5}, {{c, 1, "c"}, 0.1, 5}]
```

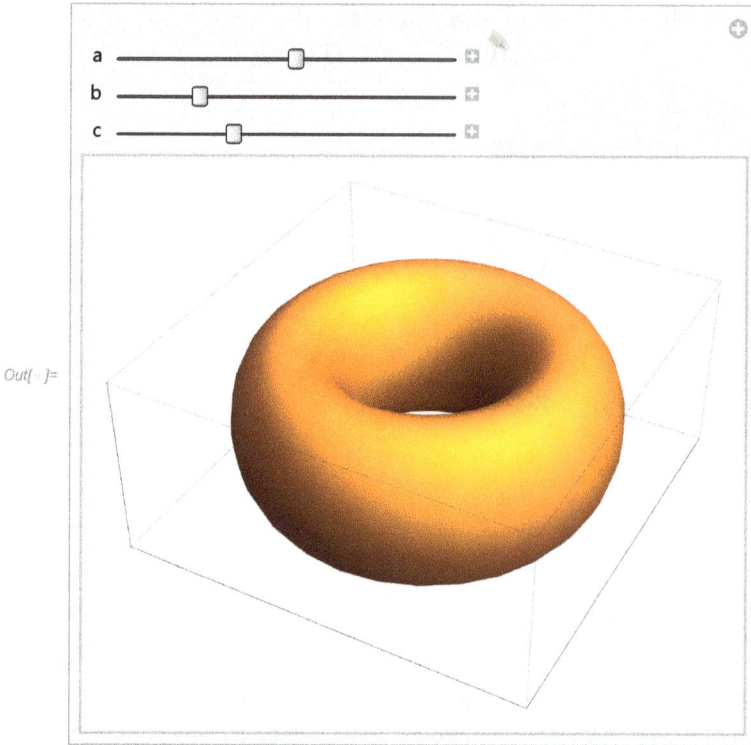

Out[ ]=

Figure 4.3

Classifying topological spaces, or even manifolds, up to homeomorphism is usually a very difficult (and in general impossible) task. Because of this, topologists often settle for something less ambitious. One approach is to replace classification up to homeomorphism with one up to *homotopy equivalence*. Two spaces X and Y are *homotopy equivalent* if there exist continuous maps $f : X \to Y$ and $g : Y \to X$ such that the composite maps $f \circ g : Y \to Y$ and $g \circ f : X \to X$ are homotopic to the identity map (two maps f, $g : X \to Y$ are said to be *homotopic* if there exists a continuous map $F : [0,1] \times X \to Y$ such that $F(0,x) = f(x)$ and $F(1,x) = g(x)$). For example, the disk (but not the circle!) is homotopy equivalent to a point (contractible). Of course, the disk is not homeomorphic to a point.

```
In[·]:= Manipulate[Graphics[{Opacity[0.5],
        Red, Disk[{0, 0}, r], Blue, Point[{0, 0}]},
        PlotRange -> {{-1, 1}, {-1, 1}}], {r, 1, 0.01},
        SaveDefinitions -> True]
```

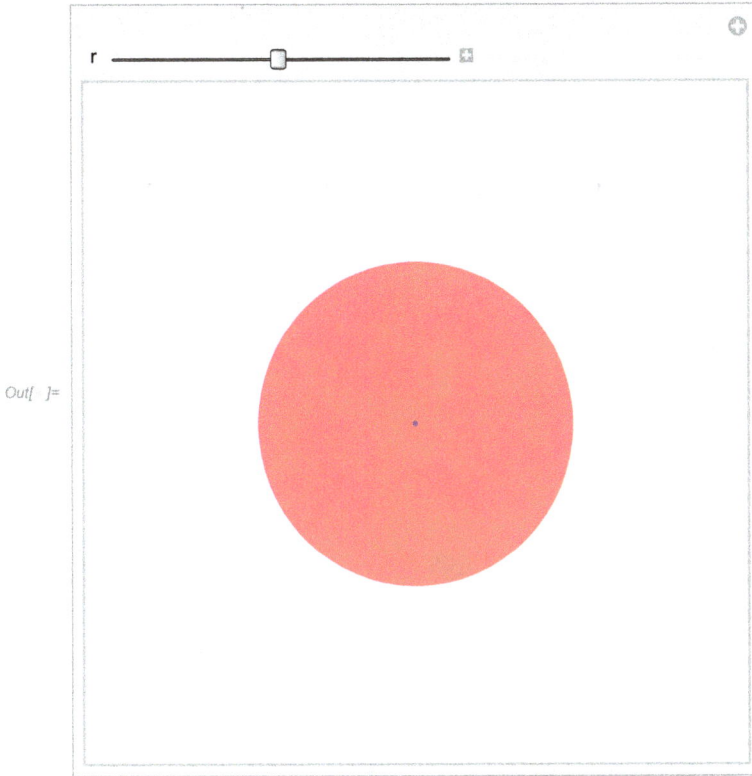

Figure 4.4

A circle with an interval attached can be continuously deformed (by a homotopy equivalence) onto a circle:

```
In[·]:= Manipulate[Graphics[{Red, Circle[{0, 0}, 1],
        Line[{{1/Sqrt[2], 1/Sqrt[2]}, {t, t}}]},
        PlotRange -> {{-2, 2}, {-2, 2}}],
        {t, 1/Sqrt[2] + 1, 1/Sqrt[2]},
        SaveDefinitions -> True]
```

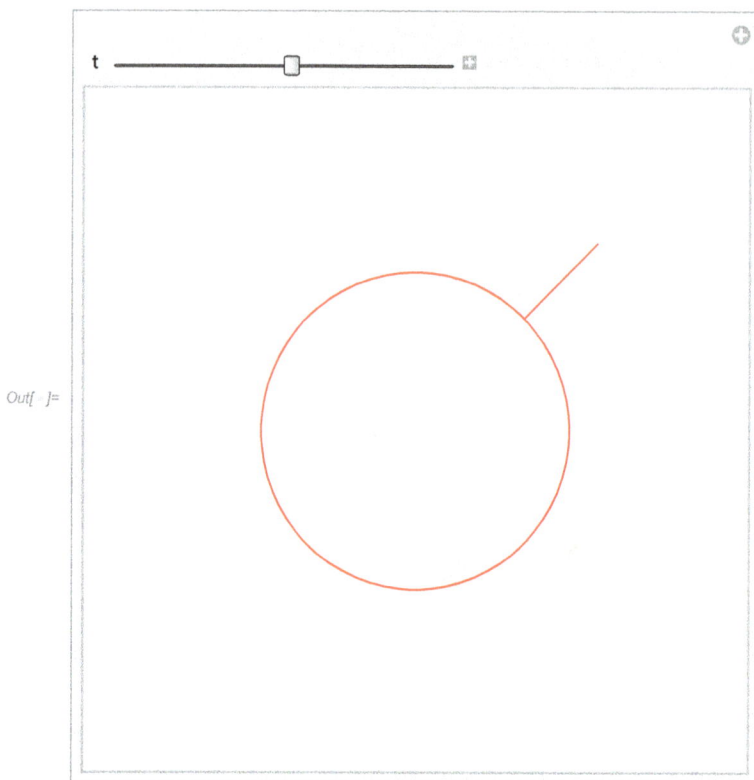

Out[ ]=

Figure 4.5

4.1.3 Homology and the Euler characteristic

The sphere produced by **Mathematica**®'s plotting function, such as `Parametric-Plot3D`, is actually a polyhedron made up of regular polygons. In fact, it is another kind of manifold called a PL-manifold. We will not discuss this here but for surfaces the concepts of smooth, topological, and PL-manifold are equivalent (in a certain sense). This is not true in general. On the other hand, **Mathematica**®'s graphics primitive `Sphere` produces what we can think of as a smooth sphere. The two are, of course, homeomorphic.

```
In[·]:= Grid[{{ParametricPlot3D[{Cos[u]*Cos[v],
        Cos[v]*Sin[u], Sin[v]}, {u, 0, 2*Pi},
        {v, -Pi/2, Pi/2}, PlotRange -> All, Ticks -> False],
        Graphics3D[Sphere[]]}}]
```

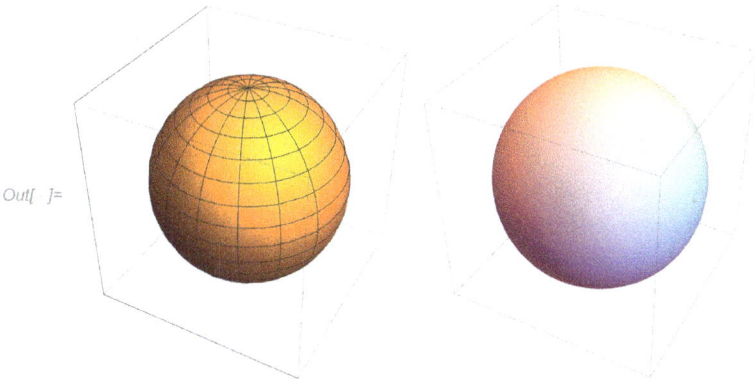

Figure 4.6

A smooth manifold can be "decomposed" into polygons in lots of ways. However, there are "invariants" that can be computed from these polyhedral decompositions, such as homology and cohomology groups. They are "invariants" in the sense that they depend only on the homotopy class of the manifold, not on the particular polyhedral structure. The simplest and the most famous such invariant is the *Euler characteristic*, which is defined as the number of all the vertices plus the number of all the edges minus the number of all the faces. This number does not depend on the decomposition and will be the same for homotopy equivalent manifolds.

The kth *Betti number* is the rank of kth homology group. Informally it refers to the number of k-dimensional holes on a topological surface. One can think of b_0 as the number of connected components, b_1 as the number of one-dimensional holes, and b_2 as the number of cavities. For instance, for a sphere we have $b_0 = 1$, $b_1 = 0$, $b_2 = 1$. On the other hand, for a torus we have $b_0 = 1$, $b_1 = 2$, and $b_2 = 1$. The *Euler characteristic* for a surface is then $b_0 - b_1 + b_2$.

4.1.4 Compact orientable surfaces

A very special feature of surfaces is that for 2-manifolds the classifications up to diffeo-morphism, homeomorphisms, homotopy, and homology coincide. More precisely, all 2-manifolds are of two kinds, orientable and nonorientable. Nonorientable compact surfaces will be left for later. Now we only note that they cannot be embedded in $\mathbb{R}^3$ without self-intersections.

A *smooth orientable surface* is one that admits a choice of a normal vector field as shown below for the case of the torus (note that here we choose a different parametrization of the torus then before). There are, of course, always two such choices, and each is said to constitute a choice of orientation.

In[·]:= tor[u_, v_] := {(4 + Cos[u])*Sin[v],
        Sin[u], (4 + Cos[u])*Cos[v]};
        nr[f_][{x_, y_}] := Normalize[Cross @@
        Transpose[D[f[u, v], {{v, u}}]]] /.{u -> x, v -> y};
        normVect[f_][{u_, v_}] :=Arrow[Tube[{f[u, v], f[u, v]
        + nr[f][{u, v}]}, 0.02]];
        ttr = ParametricPlot3D[tor[u, v],
        {u, -Pi, Pi}, {v, -Pi, Pi}, Axes -> False,
        Mesh -> None, ColorFunction ->
        (Directive[Opacity[0.4], Green] &)];

In[·]:= Manipulate[Show[ttr, Graphics3D[
        {Thick, Red, Arrowheads[Small], normVect[tor][p]}],
        PlotRange -> {{-6, 6}, {-6, 6}, {-6, 6}}],
        {{p, {0, 0}}, {-Pi, -Pi}, {Pi, Pi},
        Slider2D}, SaveDefinitions -> True]

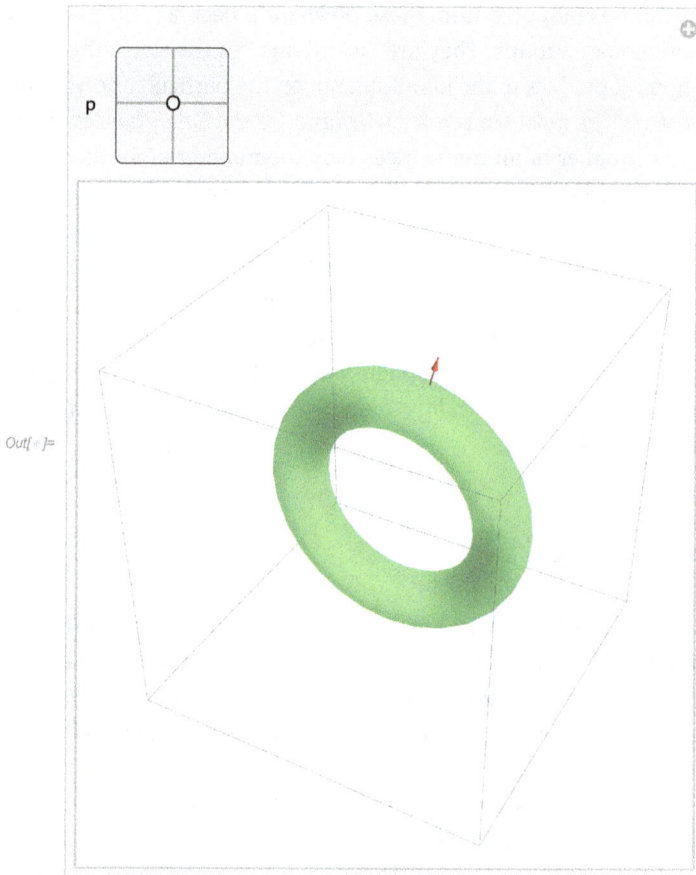

Figure 4.7

It turns out that, up to homeomorphism, all (compact, closed) orientable surfaces can all be constructed as follows. We start with a sphere and slice off two disks from it:

```
In[·]:= g1 = ContourPlot3D[x^2 + y^2 + z^2 == 1,
       {x, -1, 1}, {y, -1, 1}, {z, -1, 1},
       ColorFunction -> ({Opacity[1], Green} & ),
       RegionFunction -> (#1^2 + #2^2 > 1/16 & ),
       AxesLabel -> {"x", "y", "z"}, Ticks ->
       False, Mesh -> False, Boxed -> False, Axes -> False];
```

```
In[·]:= g2 = Graphics3D[Cylinder[{{0, 0, -1},
       {0, 0, 1}}, 1/8], Boxed -> False];
```

```
In[·]:= GraphicsGrid[{{g1, g2}}]
```

Out[·]=

Figure 4.8

```
In[·]:= Manipulate[Show[Graphics3D[{Red,
       PointSize[0.02], Point[{2, 2, 0}]}],
       ParametricPlot3D[Evaluate[RotationTransform[
       v, {0, 1, 0}, {2, 2, 0}][{Cos[u]/4, Sin[u]/4,
       Sqrt[15]/4}]], {u, 0, 2*Pi}, {v, Pi/300, b},
       Mesh -> False, Ticks -> False, AxesLabel ->
       {"x", "y", "z"}, ColorFunction -> ({Opacity[1],
       Green} & )], g1, PlotRange -> All, Boxed ->
       False, Axes -> False], {{b, Pi/12, "b"}, Pi/12,
       2*Pi - Pi/5}, SaveDefinitions -> True]
```

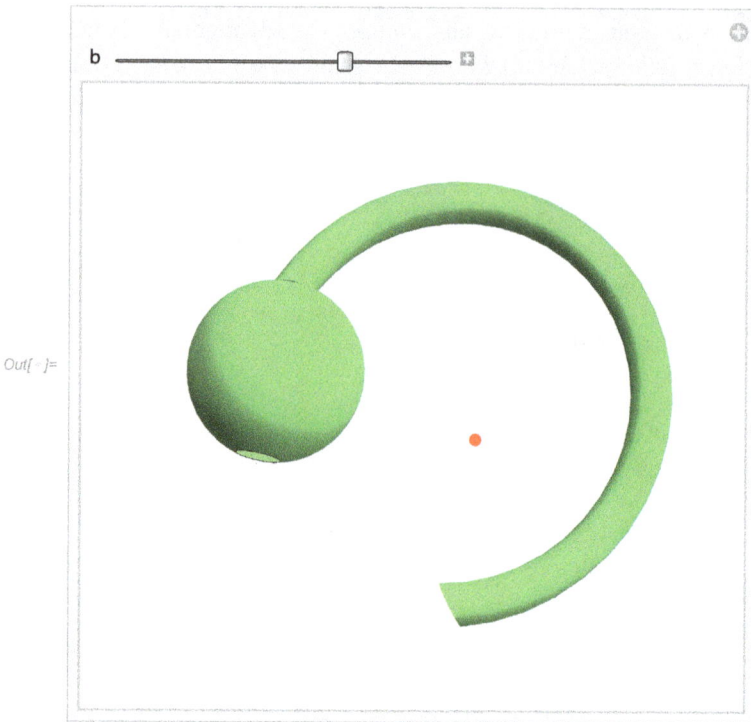

Figure 4.9

Here we use the function `RotationTransform[v,w,p]` which gives a 3D rotation around the axis w anchored at the point p. Essentially we cut out a disc on the sphere and rotate the boundary circle around the red point and around the y-axis. In our case we have

$In[\cdot]:=$ `RotationTransform[v, {0, 1, 0}, {2, 2, 0}]`
`[{Cos[u]/4, Sin[u]/4, Sqrt[15]/4}]`

$Out[\cdot]:= \left\{-2(-1 + \cos[v]) + \frac{1}{4}\cos[u]\cos[v] + \frac{1}{4}\sqrt{15}\sin[v], \right.$
$\left. \frac{\sin[u]}{4}, \frac{1}{4}\sqrt{15}\cos[v] + 2\sin[v] - \frac{1}{4}\cos[u]\sin[v]\right\}$

In the illustration above the cylinder actually goes inside the sphere, but this is not visible.

The surface obtained in this way is not, strictly speaking, smooth because of the sharp edges at the places where the cylinder is attached. However, there is a natural way to smooth these edges, which we will ignore here. Continuing this process of "attaching cylinders", we can obtain every possible compact orientable 2-manifold.

An equivalent approach is to directly glue together compact surfaces via connected sums. The *connected sum* of two manifolds (without boundary) is obtained by cutting a disk out of each (thus creating two manifolds with boundary) and attaching them along the common boundary (strictly speaking, we need some homeomorphism (or diffeomorphism) between the boundaries that are being glued to make the identification. Again, the surface that is obtained in this way is not strictly smooth but can be "smoothed". To illustrate this concept below, we draw the torus and a sphere next to each other without cutting discs for simplicity:

```
In[·]:= Show[{ParametricPlot3D[ttorus[3/2, 1, 1]
       [u, v], {u, 0, 2*Pi}, {v, 0, 2*Pi}, Mesh ->
       False, Boxed -> False, Axes -> False],
       ParametricPlot3D[{4 + 2*Cos[u]*Sin[v],
       2*Sin[u]*Sin[v], 2*Cos[v]}, {u, 0, 2*Pi},
       {v, 0, Pi}, Mesh -> False]}]
```

Out[ ]=

Figure 4.10

The space which we obtain has one hole and is called a torus (up to homotopy).

It turns out that every compact, orientable surface is homeomorphic (and if it is smooth, diffeomorphic) to a sphere with g cylinders attached. Thus, any compact, orientable, closed surface is a sphere with g cylinders attached or a torus with $g - 1$ cylinders. The number of cylinders g is called the *genus* of the surface and the *Euler characteristic* is $2 - 2g$. Hence, if we know the genus or the Euler characteristic of a closed orientable surface, then we know all about its topology and even about its differentiable structure. This is not true in general, which is why the general classification problem is so hard.

One can also show that all compact nonorientable surfaces can be obtained by removing a disk from a sphere and attaching Möbius bands along their boundaries (which are circles).

4.2 Elements of the Morse theory

4.2.1 The torus and the height function

The torus is usually described as the closed surface obtained by identifying the opposite edges of a rectangle as we saw earlier. We will need here, however, a concrete embedding of a torus into $\mathbb{R}^3$. Such an embedding can be obtained by rotating a circle about an axis in the plane of the circle which does not cross the circle. First we define a function which embeds a circle into $\mathbb{R}^3$:

In[·]:= `circle[u_] := {0, Sin[u], Cos[u] + 4}`

We can picture this circle in relation to the coordinates by using the function `ParametricPlot3D`:

In[·]:= `ParametricPlot3D[circle[t], {t, -Pi, Pi},`
    `ColorFunction -> (Red & ), AxesLabel -> {"x", "y",`
    `"z"}, ImageSize -> Small, Ticks -> False]`

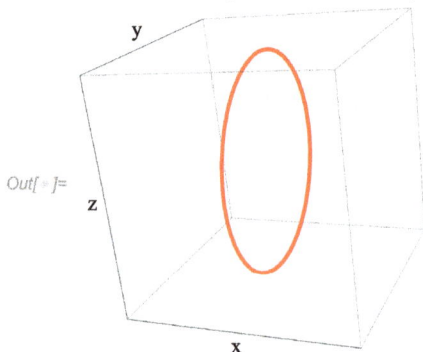

Out[·]=

Figure 4.11

We now define a function tor that rotates a point on the above circle with parameter u about the y-axis through an angle v. We use **Mathematica**®'s built-in `RotationTransform` function:

In[·]:= `tor[u_, v_] = RotationTransform[v,`
    `{0, 1, 0}][circle[u]]`
Out[·]:= `{(4 + Cos[u]) Sin[v], Sin[u], (4 + Cos[u]) Cos[v]}`

We will be frequently drawing a torus, so we define a function that will do it with various specified parameters and options. First we define some options for our function:

In[·]:= `Options[torus] = {ColorFunction ->`
    `Automatic, Mesh -> Automatic, Axes ->`
    `False, RegionFunction -> Automatic, Boxed ->`
    `True, ImageSize -> Small};`

Now we define a function to draw a torus:

```
In[·]:= torus[a_: - Pi, b_: Pi, c_: - Pi, d_: Pi,
        OptionsPattern[]] := ParametricPlot3D[
        tor[u, v], {v, c, d}, {u, a, b},
        ColorFunction -> OptionValue[ColorFunction],
        Mesh -> OptionValue[Mesh], Axes ->
        OptionValue[Axes], Boxed -> OptionValue[Boxed],
        ImageSize -> OptionValue[ImageSize],
        Evaluated -> True]
```

```
In[·]:= uaxis[v_] := ParametricPlot3D[
        Evaluate[tor[u, v]], {u, -Pi, Pi}];
```

The next dynamic illustration shows how the torus is generated by rotating a circle about an axis:

```
In[·]:= Manipulate[Show[torus[-Pi, Pi, 0, v,
        Mesh -> None,  ColorFunction -> (Directive[
        Opacity[0.4], Green] & )], uaxis[v],
        BoxRatios -> {1, 2, 2}, PlotRange ->
        {{-6, 6}, {-6, 6}, {-6, 6}}, Boxed ->
        False], {v, 0.01, 2*Pi}, SaveDefinitions -> True]
```

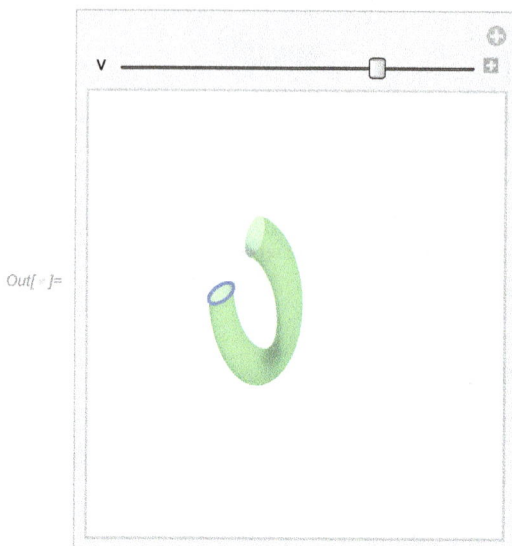

Figure 4.12

Given a torus in $\mathbb{R}^3$ as in the picture above, we can consider a "height function" on it, obtained by restricting the standard height on $\mathbb{R}^3$, i.e., the z-coordinate. In each local coordinate patch on the torus, this "height function" can be expressed in local coordinates and one can perform with it all the usual operations of calculus of several variables. In general, a surface (or a differentiable manifold) such as a torus cannot be covered by a single coordinate "patch", but requires a number of intersecting coordinate patches. We will be interested in studying those aspects of calculus on a manifold that are independent of the choice of particular coordinate patch. An example of such a concept is a critical point of a function, that is, a point in which the derivative vanishes. It is easy to show that the definition is independent of the choice of the coordinate system.

In our local coordinate system, the height function is given as the value of the last coordinate:

```
In[·]:= height[u_, v_] = tor[u, v][[3]]
Out[·]:= (4 + Cos[u]) Cos[v]
```

The values of our "height" function lie in the range between –5 and 5. If we wanted the height function to take only positive values, we could use `TranslationTransform` to move the torus up by 5 units. We will now study its critical points.

4.2.2 Critical points of the height function

The *critical points* of a function are the points where the gradient takes the value 0, in other words, where all the partial derivatives vanish. It is easy to show that the notion of a critical point is independent of the local coordinate system. Mathematica® has a built-in function Grad (which in StandardForm and TraditionalForm looks like the usual mathematical symbols for this operation):

In[·]:= Grad[f[x, y], {x, y}] == 0
Out[·]:= $\{f^{(1,0)}[x, y], f^{(0,1)}[x,y]\} == 0$

Next we look for all critical points of the height function on the torus. We can use the functions Solve or Reduce. Solve has the advantage that it gives its answer in the form of rules, which is often convenient, but Reduce can solve a greater variety of equations (and inequalities). One way to get both advantages is to use Solve with the option: Method → Reduce.

In[·]:= solrules = Solve[Grad[height[u, v],
        {u, v}] == 0 && Inequality[-Pi, Less, u,
        LessEqual, Pi] && Inequality[-Pi, Less, v,
        LessEqual, Pi], {u, v}, Method -> Reduce]
Out[·]:= {{u → 0, v → 0}, {u → 0, v → π}, {u → π, v → 0},
        {u → π, v → π}}

We would like to sort the solutions according to the height (from the lowest to the highest), so we use the command Sort with a suitable sorting function as the second argument:

In[·]:= solrules = Sort[solrules, (height[u, v] /. #1)
        < (height[u, v] /. #2) &];

The coordinates of the critical points are

In[·]:= sols = {u, v} /. solrules
Out[·]:= {{0, π}, {π, π}, {π, 0}, {0, 0}}

Now let us calculate the heights at the critical points (the critical values of the height function):

In[·]:= heights = Apply[height, sols, {1}]
Out[·]:= {-5, -3, 3, 5}

Finally, we show some pictures of the torus and the critical points.

In[·]:= criticalpoints = Graphics3D[({Red,
        PointSize[0.02], Point[#1]} &) /@
        (tor[u, v] /. solrules), Boxed -> False];

In[·]:= Show[{torus[-Pi, Pi, -Pi, Pi, Mesh ->
    None, ColorFunction -> (Directive[Opacity[0.3],
    Green] &)]}, criticalpoints, Boxed -> True,
    PlotRange -> All, Axes -> False,
    ImageSize -> Small]

Out[·]=

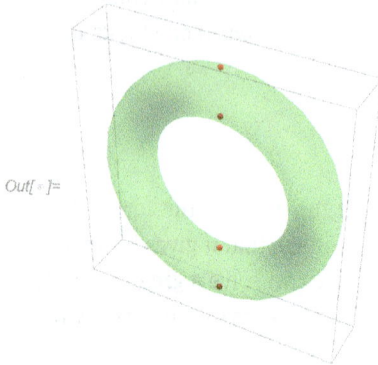

Figure 4.13

4.2.3 The Hessian and the index of a critical point

If P is a critical point of a smooth function f, we define the *Hessian* of f at P as the symmetric bilinear form on the tangent space at P, which in a local coordinate system can be represented as the matrix of second order derivatives

$$\left(\frac{\partial^2 f}{\partial x_i \partial x_j}(P) \right).$$

This is defined for a smooth function on a manifold of arbitrary dimension d, in which case the Hessian is a $d \times d$ matrix. In the case of surfaces $d = 2$. This definition (the matrix) depends on the choice of coordinates at P, but the corresponding symmetric bilinear form (on the tangent space) and its determinant is independent of the choice of local coordinates. In Mathematica® we can define the Hessian as:

In[·]:= hessian[f_, vars_List] := D[f, {vars, 2}]

A critical point is said to be *nondegenerate* if its Hessian has maximal rank, i. e., its null space is zero. One can show that all nondegenerate critical points are isolated.

We now compute the Hessian matrices at all our critical points:

In[·]:= MatrixForm /@ (hessians =
    hessian[height[u, v], {u, v}] /. solrules)

$$\textit{Out[·]:=} \left\{ \begin{pmatrix} 1 & 0 \\ 0 & 5 \end{pmatrix}, \begin{pmatrix} -1 & 0 \\ 0 & 3 \end{pmatrix}, \begin{pmatrix} 1 & 0 \\ 0 & -3 \end{pmatrix}, \begin{pmatrix} -1 & 0 \\ 0 & -5 \end{pmatrix} \right\}$$

It is easy to check with Mathematica® that the critical points are nondegenerate:

In[·]:= `Map[NullSpace, hessians]`
Out[·]:= `{{}, {}, {}, {}}`

A function all of whose critical points are isolated and nondegenerate is called a *Morse function.* So the height function of the torus given by the z-coordinate is indeed a Morse function.

The *index of a nondegenerate critical point* is defined as the number of negative eigenvalues of the Hessian. It can be defined in Mathematica® as follows:

In[·]:= `index[f_, x_List, a_List] := Count[Eigenvalues[`
         `hessian[f, x] /. Thread[x -> a]], _?Negative]`

Here are the indices of our four critical points:

In[·]:= `indices = (index[height[x, y],`
         `{x, y}, #1] & ) /@ sols`
Out[·]:= `{0, 1, 1, 2}`

When studying the topology of a manifold by means of a Morse function, it is often useful to consider the *Morse polynomial,* which is a polynomial in a variable t, whose t^i-coefficient is the number of critical points of index i. The Mathematica® definition is

In[·]:= `MorsePolynomial[ind_List, t_] := Sum[Count[`
         `ind, i]*t^i, {i, Min[ind], Max[ind]}]`

For our torus, we have

In[·]:= `MorsePolynomial[indices, t]`
Out[·]:= `1 + 2 t + t`^{2}

The *Poincare polynomial* of a topological space is defined in an analogous way, but has as its coefficients the Betti numbers:

In[·]:= `PoincarePolynomial[b_List, t_] :=`
         `b . Table[t^i, {i, 0, Length[b] - 1}]`

A surface has three Betti numbers b_0, b_1, b_2, and the Poincare polynomial has the form

In[·]:= `PoincarePolynomial[{b0, b1, b2}, t]`

Out[·]:= `b0 + b1 t + b2 t`^{2}

In the case of the torus, we have

In[·]:= `PoincarePolynomial[{1, 2, 1}, t]`
Out[·]:= `1 + 2 t + t`^{2}

Thus in this case the Poincare polynomial is the same as the Morse polynomial. In general we have the following relation: MorsePolynomial(indexes, t)–PoincarePolynomial (Betti numbers, t) $= (t + 1)Q(t)$, where $Q(t)$ is some polynomial. Substituting $t = -1$ in this equation, we see that the value of the Morse polynomial at $t = -1$ is the Euler characteristic:

>*In[·]:=* EC[indices_] := MorsePolynomial[indices, -1]

In the case of the torus,

>*In[·]:=* EC[indices]
>*Out[·]:=* 0

All these formulas are true for manifolds of arbitrary dimension d, in which case there are $d + 1$ Betti numbers.

From the above formula relating the Morse and Poincare polynomials, one can deduce the following useful result.

Proposition 1. *Let $f : M \to \mathbb{R}$ be a Morse function on a compact manifold M. If the Morse polynomial of f contains only even powers, then it coincides with the Poincare polynomial (and hence we can compute the Betti numbers).*

4.2.4 Brief survey of Morse theory

A basic tool of Morse theory is the *Morse lemma*. It asserts that a Morse function can be expressed in local coordinates in a neighborhood of a nondegenerate critical point P in one of the forms:

$$\{f(P) - x^2 - y^2, f(P) - x^2 + y^2, f(P) + x^2 - y^2, f(P) + x^2 + y^2\},$$

where the number of minus signs is equal to the index. This implies that the level curves of the height function of the torus look like either circles or hyperbolas near the critical points, which can be seen in the interactive demonstration below.

For each real number a, let the *sublevel set* be defined as

$$M_a = f^{-1}((-\infty, a]) = \{x \in M \mid f(x) \le a\}.$$

We observe the evolution of M_a, for M being our torus and f the height function. We see that if there are no critical points between a and b, the spaces (manifolds with boundary) M_a and M_b are diffeomorphic. A diffeomorphism can be defined by "pushing down" M_b to M_a along the lines of the "gradient flow" of the height function (we shall not consider this here but see [4] and [8]). However, if there is a singular value $c = f(x)$, with $a < c < b$, where x is a critical point of f, M_b and M_a have a different homotopy type (and hence are not diffeomorphic). In fact, one can describe precisely how M_b is obtained from M_a by attaching a "handle", whose type is determined by the index of the critical point x. Thus a given Morse function f on a closed (i. e., compact, without boundary) manifold M determines a "handle decomposition" of the manifold. In general, of course, the information provided by a Morse function and thus the "handle decomposition" of a manifold is not sufficient to determine it up to a homeomorphism, but for surfaces this is true.

In[·]:= `Manipulate[Show[ParametricPlot3D[`
`{(4 + Cos[u])*Sin[v], Sin[u], (4 + Cos[u])*Cos[v]},`
`{u, -Pi, Pi}, {v, -Pi, Pi}, Axes -> False,`
`Mesh -> None, ColorFunction ->`
`(Directive[Opacity[0.4], Green] & ),`
`RegionFunction -> (If[tt, True, #3 < t] & )],`
`If[ss, Plot3D[t, {x, -5, 5}, {y, -5, 5},`
`ColorFunction -> ({Opacity[0.3], Blue} & ),`
`Mesh -> None], Graphics3D[{}]],`
`PlotRange -> {{-5, 5}, {-5, 5}, {-5, 5}}],`
`{{t, -5, "height"}, -5, 5, Appearance -> "Labeled"},`
`{{tt, True, "Show full torus"}, {True, False}},`
`{{ss, False, "Show plane"}, {True, False}},`
`Bookmarks -> {"first" :> (t = -5 + 0.01),`
`"second" :> (t = -3 + 0.001), "third" :>`
`(t = 3 + 0.001), "fourth" :> (t = 5)},`
`SaveDefinitions -> True]`

Out[·]=

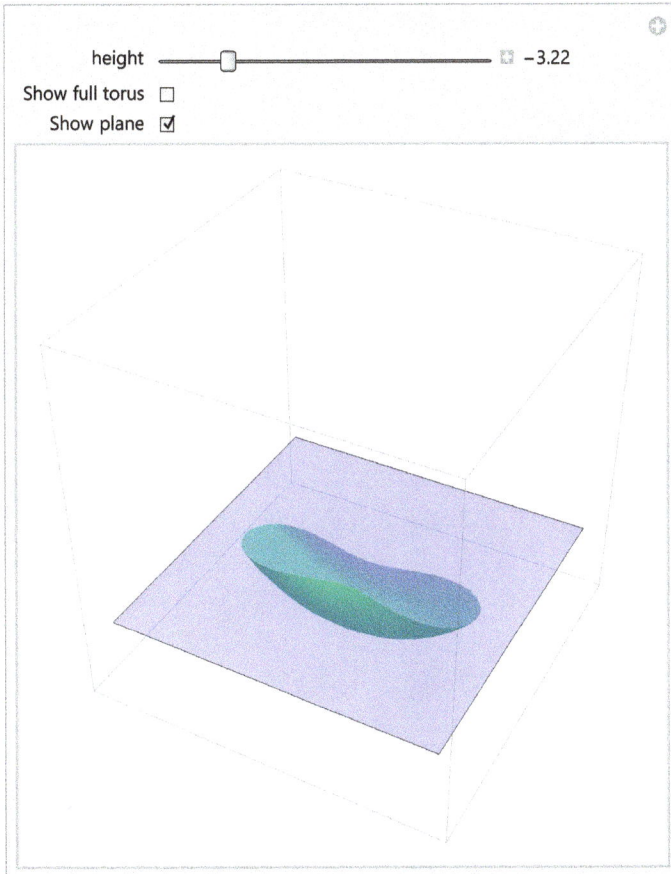

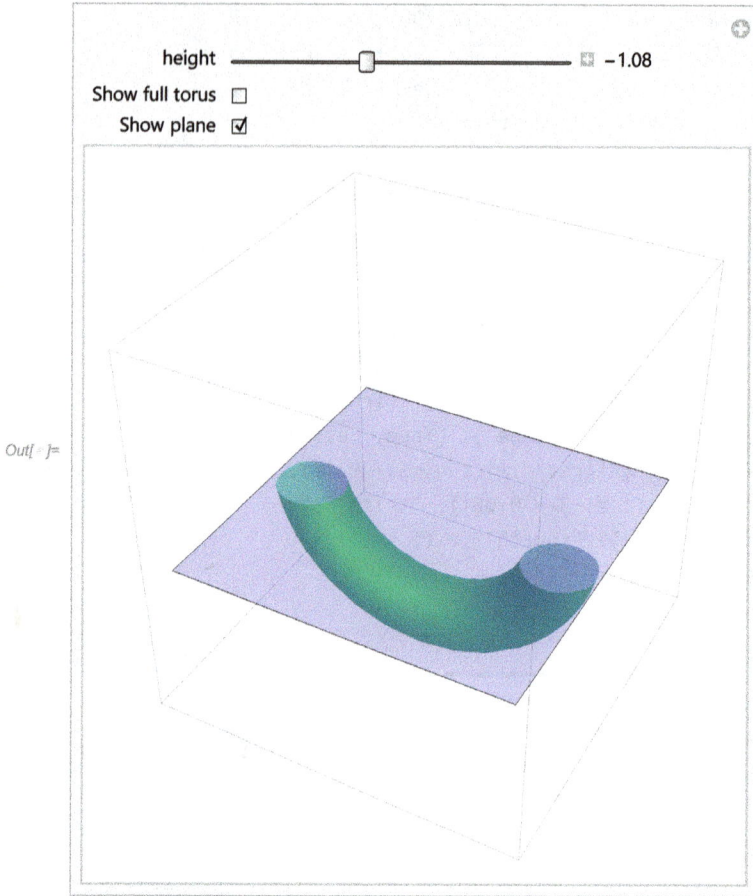

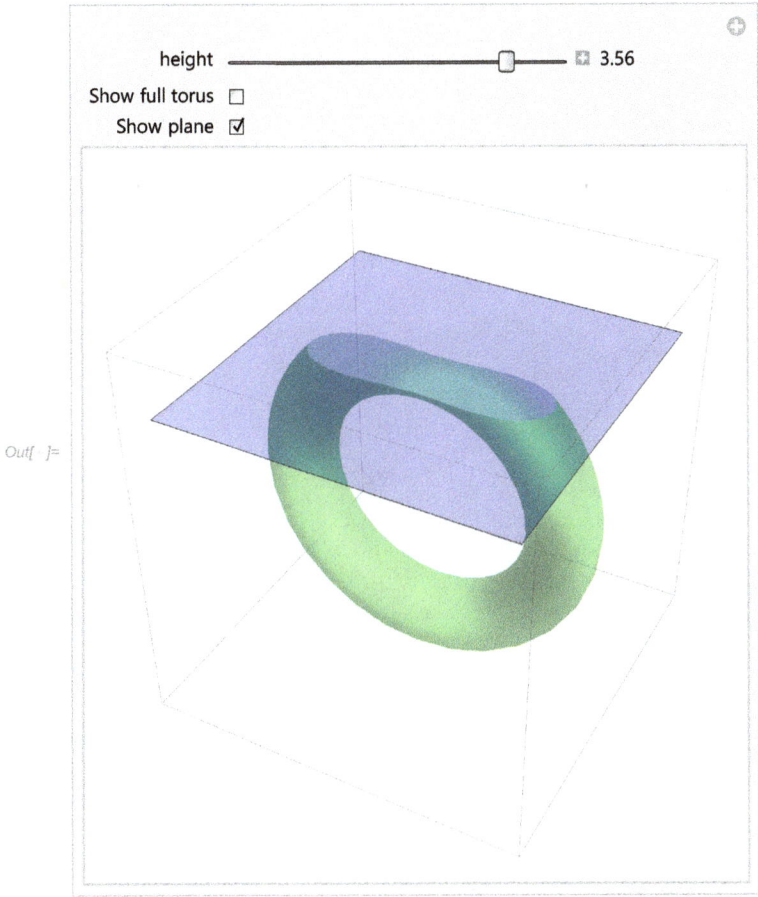

Out[ ]=

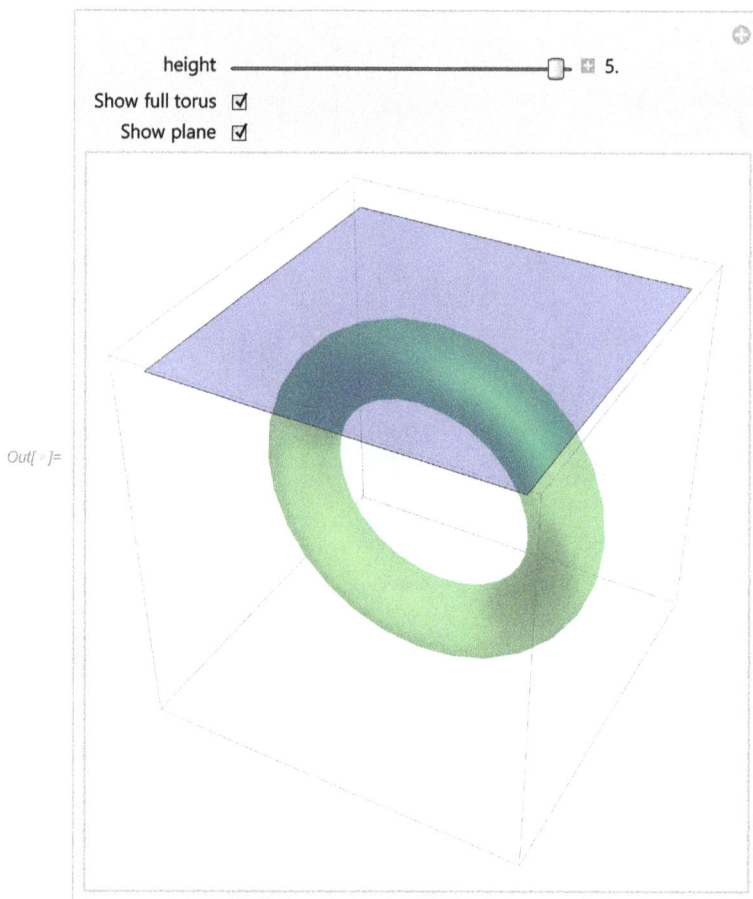

Out[ ]=

Figure 4.14

4.2.5 Nonorientable surfaces

So far we have considered only orientable surfaces, for which we can choose consistently over the whole surface an "inner" or "outer" normal vector field (i. e., a family of vectors attached at the points of the surface and perpendicular to the tangent planes at these points. However, Morse theory can also be applied in exactly the same way to nonorientable surfaces (given by parametric representation). Similar computations as above can be performed for nonorientable surfaces, for instance, for the *Klein bottle*. Although the Klein bottle cannot be embedded in $\mathbb{R}^3$ without self-intersections, one can choose an embedding such that the height function is a Morse function. The computation of indexes and the Euler characteristic proceeds as before. We include only the results of computations without any additional explanation.

```
In[·]:= klein[u_, v_] := {(2 + Cos[v/2]*Sin[u] -
        Sin[v/2]*Sin[2*u])*Cos[v], (2 + Cos[v/2]*Sin[u] -
        Sin[v/2]*Sin[2*u])*Sin[v],   Sin[v/2]*Sin[u] +
        Cos[v/2]*Sin[2*u]}
```

Out[·]=

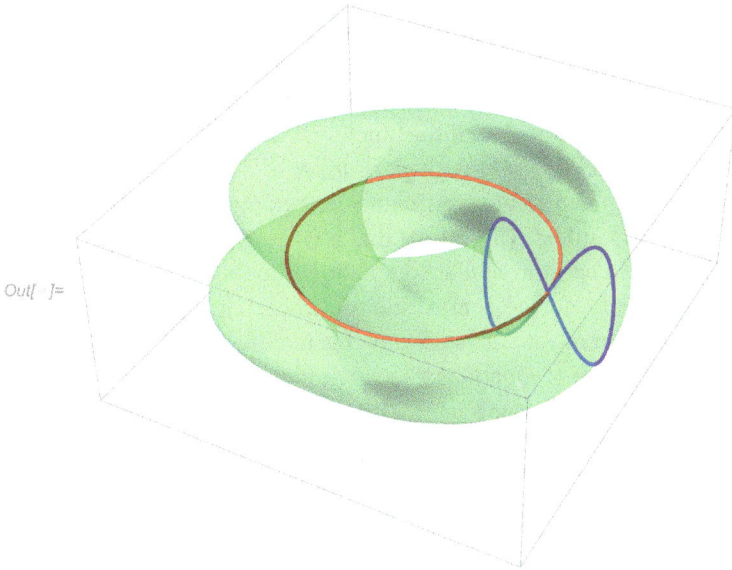

Figure 4.15

Out[·]=

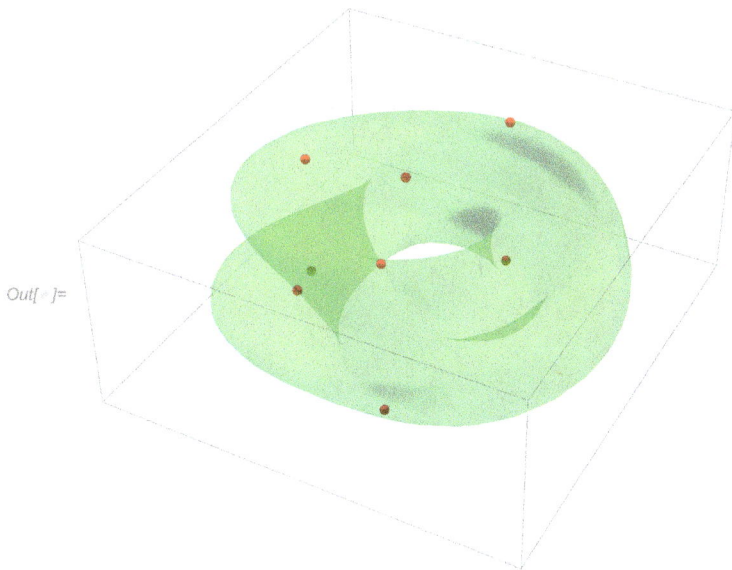

Figure 4.16

The Morse polynomial is $2 + 4t + 2t^2$, and the Euler characteristic is 0.

4.3 Surfaces defined implicitly

4.3.1 Elements of real algebraic geometry

Above we made use of parametric representation of the torus to obtain local coordinates which allowed us to compute the coordinates of singular points and their indexes. Many important surfaces which appear in geometry, topology, and applications are not given in terms of parametric representation (or local coordinate patches) but as the set of zeros of some smooth function. If the function is an irreducible polynomial, the surface is called an *irreducible algebraic (hyper)surface* [1]. By the implicit function theorem, if the points of this zero set are nonsingular (the gradient is not zero), the set of zeros will be a smooth manifold. This means that it is always possible to choose a local coordinate system and perform similar computations to those above.

It is often convenient to do the computations without choosing coordinates, purely in terms of the system of equations defining the manifold. In the real case we can actually replace the system of equations by a single equation $f_1^2 + f_2^2 + \cdots + f_m^2 = 0$. However, **Mathematica**®'s function ContourPlot3D will not be able to plot $f(x,y,z)^2 = 0$, although it can plot $f(x,y,z) = 0$. For example,

In[·]:= ContourPlot3D[x^2 + y^2 + z^2 - 1 == 0,
        {x, -1, 1}, {y, -1, 1}, {z, -1, 1}]

Out[·]=

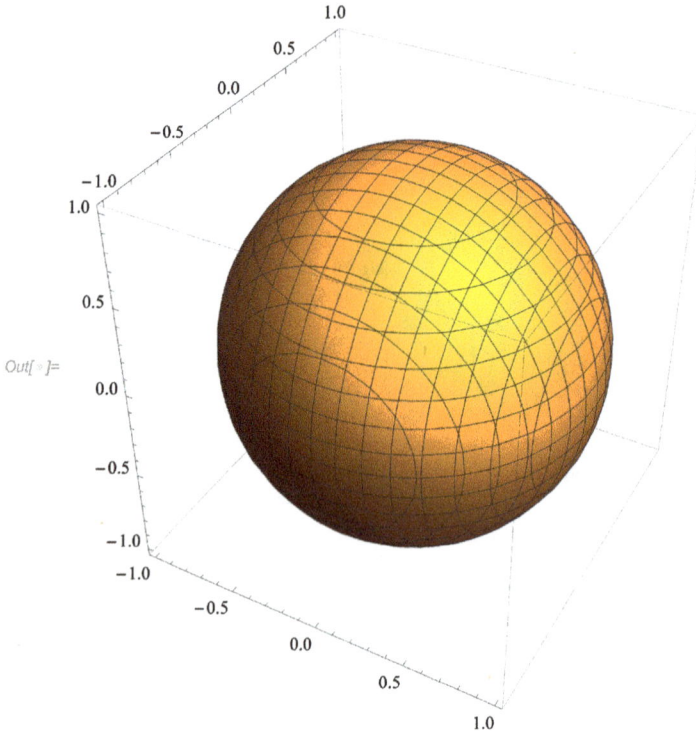

Figure 4.17

but

In[·]:= ContourPlot3D[(x^2 + y^2 + z^2 - 1)^2 == 0,
     {x, -1, 1}, {y, -1, 1}, {z, -1, 1}]

produces the empty box. However, we can still see an approximation of the surface by plotting

In[·]:= ContourPlot3D[(x^2 + y^2 + z^2 - 1)^2 == 0.01,
     {x, -2, 2}, {y, -2, 2}, {z, -2, 2}]

Out[ ]=

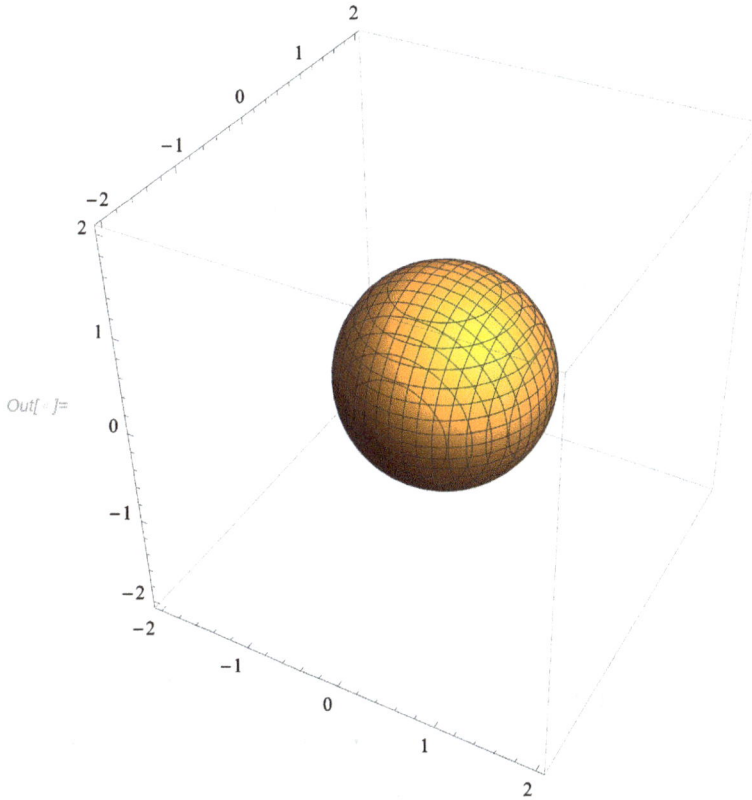

Figure 4.18

We can use the same method to find approximately the set of points that is the intersection of two surfaces, e. g.,

In[·]:= ContourPlot3D[(x + y + z)^2 +
     (x - y + z)^2 == 0.01, {x, -2, 2},
     {y, -2, 2}, {z, -2, 2}]

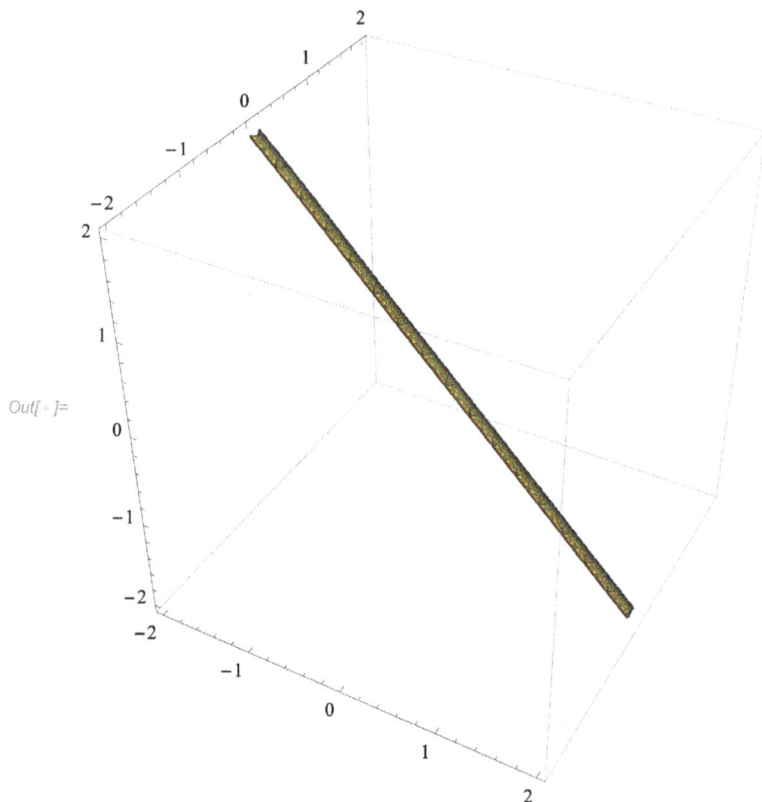

Figure 4.19

This gives a slightly thickened line. Note that in general ContourPlot and Contour-Plot3D cannot draw contours of codimension greater than 1 (that is, of dimension less than 1 in the former case and of dimension less than 2 in the latter).

When the implicit equations are algebraic, the set of solutions is called an *algebraic set* or a (real) *algebraic variety*. An algebraic subset of $\mathbb{R}^n$ is the set of common solutions of a finite set of polynomials $\{f_1, \ldots, f_m\}$ (or the ideal generated by them). Such a set is said to be nonsingular if the Jacobian matrix $(\frac{\partial f_i}{\partial x_j})$ has rank $n - m$.

The following theorem tells us that the class of nonsingular algebraic surfaces includes all compact smooth manifolds, up to diffeomorphism.

Theorem 8 (Tognoli's theorem [1, Theorem 14.1.10]). *Let $M \subset \mathbb{R}^n$ be a compact C^∞-submanifold. Then M is diffeomorphic to a nonsingular algebraic subset of $\mathbb{R}^{n+1}$.*

It is not known whether $n+1$ can be in general replaced by n. However, it is known that we can represent all closed, orientable surfaces by nonsingular algebraic sets in $\mathbb{R}^3$. There is an interesting way to do so in the case of a surface represented in parametric form as above. This method is known as "implicitization" and uses a powerful algorithm known as Gröbner basis. The function GroebnerBasis finds a certain canonical basis for a specified ideal of polynomials. The result depends on the way monomials are ordered, which also depends on the ordering of variables.

Let I be an ideal in $k[x_1, \ldots, x_n]$ (the ring of polynomials in n variables with coefficients in the field k). A Gröbner basis for I for a given monomial order $>$ (e. g., lexicographic) is a finite collection of polynomials $\{g_1, \ldots, g_m\}$ with the property that for every nonzero $f \in I$ the leading term of f is divisible by the leading term of some g_i. See [3] for more information. For instance, let us take polynomials $x^2y + z^2x$ and x^3. Their leading terms (in the default lexicographic monomial ordering) are x^2y and x^3.

In[·]:= GroebnerBasis[{x^2*y + z^2*x, x^3}, {x, y, z}]
Out[·]:= $\{x\,z^4,\ x^2\,z^2,\ x^2\,y + x\,z^2,\ x^3\}$

The original set of polynomials is not a Gröbner basis because, for example, x^2z^2 belongs to this ideal since $(x^2y + z^2x)x - zx^3 = x^2z^2$ but its leading term (which is the monomial itself) is not divisible by any of the leading terms x^2y and x^3. So we see that two more polynomials are added to the Gröbner basis in our example. Any other polynomial in this ideal would have the leading term divisible by the leading term of one of these four polynomials. The advantage of the Gröbner basis over the ordinary basis is that for a Gröbner basis we have a canonical way to express any polynomial in the ideal in terms of the basis. This is done by eliminating leading terms step by step. In particular, for a linear ideal, a Gröbner basis will always take a row echelon form:

In[·]:= GroebnerBasis[{x + y + z - 1, -x + y - z + 6,
        x - 2*y + 1}, {x, y, z}]
Out[·]:= {-19 + 2 z, 5 + 2 y, 6 + x}

In order to change the monomial ordering, we need to use the option MonomialOrder, for example,

In[·]:= GroebnerBasis[{x^2*y + z^2*x, x^3}, {x, y, z},
        MonomialOrder -> DegreeReverseLexicographic]
Out[·]:= $\{x^2\,y + x\,z^2,\ x^3,\ x^2\,z^2,\ x\,z^4\}$

The method of Gröbner bases makes it possible to solve systems of polynomial equations by "elimination of variables", which can be viewed as a generalization of the Gaussian elimination method of solving systems of linear equations. For us this ability to eliminate variables will be most important. Let us consider the following example:

In[·]:= GroebnerBasis[{x - 2*u, y - 3*v,
        u^2 + v^2 - 1}, {x, y}, {u, v}]
Out[·]:= $\{-36 + 9\,x^2 + 4\,y^2\}$

We see that variables u and v have been eliminated. We can use this property to obtain a representation of the torus as an algebraic set:

In[·]:= implicitTor[x_, y_, z_] = First[GroebnerBasis[
        Join[{x, y, z} - tor[u, v], {1 - Sin[u]^2 - Cos[u]^2,
        1 - Sin[v]^2 - Cos[v]^2}], {x, y, z},
        {Cos[u], Sin[u], Cos[v], Sin[v]}]]
Out[·]:= $225 - 34 x^2 + x^4 + 30 y^2 + 2 x^2 y^2 + y^4$
        $- 34 z^2 + 2 x^2 z^2 + 2 y^2 z^2 + z^4$

Here the trigonometric functions $\sin u$, $\cos u$, $\sin v$, $\cos v$ are treated as variables which we eliminate from the system of equations describing the torus. We obtain a single polynomial equation which represents the torus as a real algebraic hypersurface in $\mathbb{R}^3$ (surface given as the set of solutions of a single real equation). Strictly speaking, this method only produces an algebraic set which contains the given parametric surface but in our case it is obvious that the algebraic hypersurface is actually the same as the one defined parametrically. We will verify this by looking at the plot:

In[·]:= tr = ContourPlot3D[implicitTor[x, y, z],
        {x, -6, 6}, {y, -2, 2}, {z, -6, 6}, Contours -> {0},
        Mesh -> None, Axes -> False, Boxed -> True,
        BoxRatios -> {1, 0.2, 1}, ColorFunction -> (Directive[
        Opacity[0.4], Green] &), ImageSize -> Small]

Outf ·]=

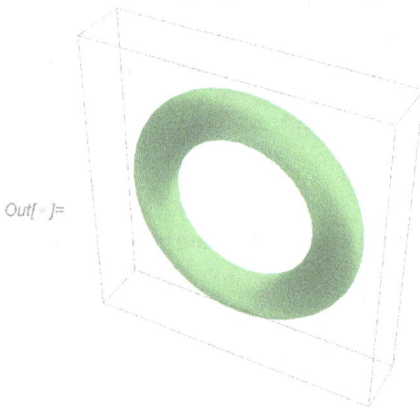

Figure 4.20

4.3.2 Morse theory on an implicit surface

The reason why implicit description is often very convenient lies in the following theorem.

Theorem 9. *Let $W \subset \mathbb{R}^n$ be a compact nonsingular algebraic hypersurface defined by the equation $f = 0$, where f is a polynomial. Then (possibly after a change of coordinates) one can find a height function h which is a Morse function whose critical points* *are the solutions of n polynomial equations*

$$f = 0, \quad \frac{\partial f}{\partial x_1} = 0, \quad \ldots, \quad \frac{\partial f}{\partial x_{n-1}} = 0.$$

Moreover, the Hessian of the height function is $-H(f)/(\frac{\partial f}{\partial x_n})$, *where* $H(f)$ *is the* $(n-1) \times$ $(n-1)$ *matrix* $(\frac{\partial^2 f}{\partial x_i \partial x_j})$.

Let us first check that our hypersurface implicitTor above is indeed nonsingular:

In[·]:= `Reduce[{implicitTor[x, y, z] == 0, Grad[implicitTor[`
       `x, y, z], {x, y, z}] == 0}, {x, y, z}, Reals]`
Out[·]:= `False`

There are no solutions, i. e., no singular points. We apply the above theorem and find the critical point:

In[·]:= `criticalPoints = {x, y, z} /. Solve[{Grad[implicitTor[`
       `x, y, z], {x, y}] == 0, implicitTor[x, y, z] == 0},`
       `{x, y, z}, Reals, Method -> Reduce]`
Out[·]:= `{{0, 0, -5}, {0, 0, -3}, {0, 0, 3}, {0, 0, 5}}`

In[·]:= `crits = {PointSize[0.02], (Point[#1] & )`
       `/@ criticalPoints};`

In[·]:= `criticalValues = criticalPoints[[All, 3]]`
Out[·]:= `{-5, -3, 3, 5}`

In[·]:= `hess = -(hessian[implicitTor[x, y, z],`
       `{x, y}]/D[implicitTor[x, y, z], z]);`

In[·]:= `Thread /@ ({x, y, z} -> #1 & ) /@`
       `{{0, 0, -5}, {0, 0, -3}, {0, 0, 3}, {0, 0, 5}}`
Out[·]:= `{{x → 0, y → 0, z → -5}, {x → 0, y → 0, z → -3},`
       `{x → 0, y → 0, z → 3}, {x → 0, y → 0, z → 5}}`

In[·]:= `hessians = hess /. %;`

In[·]:= `MatrixForm /@ hessians`
Out[·]:= $\left\{ \begin{pmatrix} \frac{1}{5} & 0 \\ 0 & 1 \end{pmatrix}, \begin{pmatrix} \frac{1}{3} & 0 \\ 0 & -1 \end{pmatrix}, \begin{pmatrix} -\frac{1}{3} & 0 \\ 0 & 1 \end{pmatrix}, \begin{pmatrix} -\frac{1}{5} & 0 \\ 0 & -1 \end{pmatrix} \right\}$

In[·]:= `(Count[Eigenvalues[#1], _?Negative] & )`
       `/@ hessians`
Out[·]:= `{0, 1, 1, 2}`

Thus we obtained the same result as earlier, but without using a parametric description (or local coordinates).

4.3.3 Example: double torus

The above method is particularly useful in situations when we do not know any parametric description of a surface. Consider the real algebraic hypersurface defined by the following equation:

In[·]:= `implicit2[x_, y_, z_] := -8 + 16*x^2 + 16*y^4 -`
`        32*y^2*z^2 + 16*z^4 + 8*y^2*z^4 - 8*z^6 + z^8`

Let us first verify that the surface is nonsingular:

In[·]:= `singularPoints1 = Solve[Join @@`
`        {Grad[implicit2[x, y, z], {x, y, z}],`
`        {implicit2[x, y, z]}} == 0,`
`        {x, y, z}, Reals, Method -> Reduce]`
Out[·]:= `{}`

Repeating computations for the torus given by `implicitTor`, we can find critical points and calculate the Morse polynomial for our surface. The Morse polynomial is equal to $1 + 4t + t^2$ and the Euler characteristic is -2. One can plot our surface and the critical points. The result is

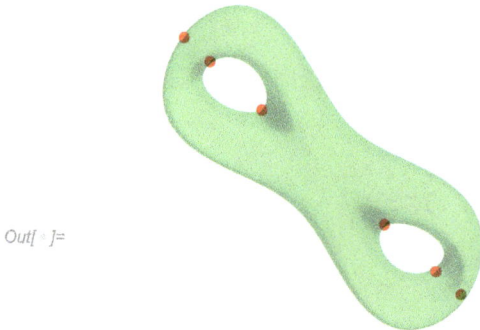

Out[·]=

Figure 4.21

It turns out that our surface is a "double torus", which has genus 2 (where genus is "the number of holes"). It is known that the Euler characteristic of an orientable surface is $2 - 2g$, where g is the genus, which shows that in our case it is indeed -2.

5 The Riemann integral of a function of *n* variables

In this chapter we discuss, without full details, two notions of integration of functions of two or more variables, Riemann and Lebesgue integration. We assume the knowledge of the basic concept of a "measure" without any further explanation. The reader should consult the recommended textbooks in the bibliography. We show how Mathematica® computes integrals over various domains in many dimensions. However, to perform such computations by hand, we need two theorems which we state without proof. Fubini's theorem replaces the problem of computing the integral in $\mathbb{R}^n$ by the problem of computing iterated integrals of functions of one variable. The change of variables theorem can often be used to reduce the problem of computing an integral over some complicated regions to the problem of computing another integral over a simpler domain (usually a rectangular box). We give examples of using both methods. Although these methods are not required if one uses Mathematica®, they are important in applications and for understanding the theory.

5.1 Riemann vs Lebesgue integration

In Volume 1 [5] we defined the Riemann integral of a bounded function on a compact (that is, closed and bounded) subset of $\mathbb{R}$. The functions for which the integral is defined are called Riemann integrable functions. They include all continuous functions but also some discontinuous ones (those for which the set of points where the function is discontinuous has measure zero). We then extended the definition of integral to certain functions that were not bounded or defined on nonclosed intervals (improper integrals). The problem with this approach is that many natural functions, even bounded ones defined on closed intervals (e. g., the Dirichlet function), are not Riemann integrable. However, in the case of functions of one variable, the simplicity of the definition of the Riemann integral and the fact it is sufficient for most applications make it the method of choice in most courses. In the case of functions of two variables, the rigorous definition of a Riemann integral is almost as complicated as that of the much more general Lebesgue integral, which is why in most texts Riemann integral is used in the one variable case and Lebesgue integral in the many variable case. Here, however, we shall not be concerned with the formal definitions (since the Lebesgue and Riemann integrals agree in all the cases when the Riemann integral is defined) and we are primarily interested in computing integrals with Mathematica®. We refer to [14] for the theory of the *Lebesgue integral*, and to [10] or [13] for the theory of *Riemann integrals* in more than one dimension.

  We will begin with the special case in which the theory is simple: *integration over a closed rectangle* (or rectangular box when the number of variables is greater than 2). In this case the definition of the Riemann integral of a function of *n* variables is completely analogous to that for a function of one variable.

https://doi.org/10.1515/9783110660395-005

Let us first recall the intuitive geometric idea behind the Riemann integration of a function of one variable. We think of the definite integral $\int_a^b f(x)\,dx$ as the area under the graph of $f(x)$. The function is Riemann integrable if this area can be approximated arbitrarily closely by a sum of areas of rectangles as illustrated below.

```
In[·]:= f[x_] := x^2
In[·]:= g1 = Plot[f[x], {x, 0, 1}];
In[·]:= g2 = With[{h = 0.01}, Graphics[Table[
        {RGBColor[RandomReal[{0, 1}, 3]],
        Rectangle[{x, 0}, {x + h, f[x]}]},
        {x, 0, 1 - h, h}], AspectRatio -> 1]];
In[·]:= Show[g1, g2, Axes -> True]
```

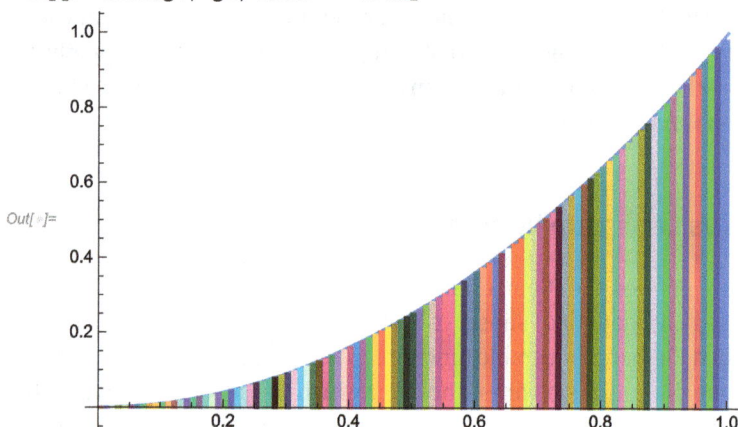

Figure 5.1

We can check numerically that the sum of areas of rectangles is a good approximation to the area under the graph, and, by increasing the number of rectangles while decreasing the width, we can make the approximation as accurate as we like:

```
In[·]:= With[{h = 0.001}, Sum[f[i*h]*h, {i, 1, 1/h}]]
Out[·]:= 0.333834
```

```
In[·]:= N[Integrate[f[x], {x, 0, 1}]]
Out[·]:= 0.333333
```

There is also another way to compute such an area which is based on approximating the area by a sum of little squares filling up almost completely the region under the graph:

```
In[·]:= g3 = With[{h = 0.01, k = 0.01},
        Graphics[Table[{RGBColor[RandomReal[
        {0, 1}, 3]], Rectangle[{x, y}, {x + h,
        y + k}]},   {x, 0, 1, h}, {y, 0, f[x], k}],
        AspectRatio -> Automatic]];
```

In[·]:= Show[g1, g3]

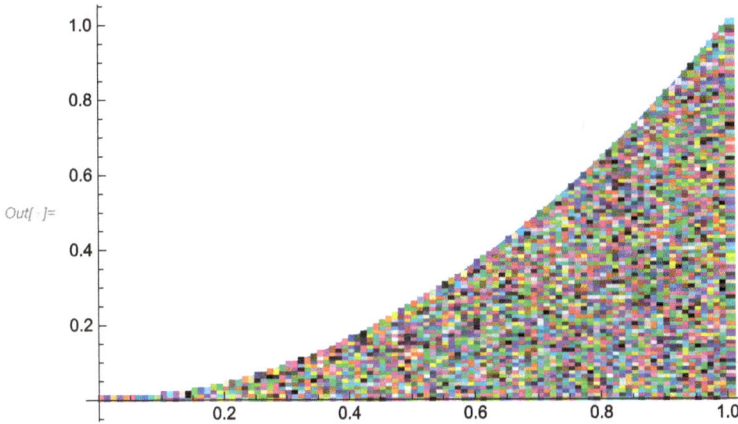

Figure 5.2

We see again that we get a good approximation:

```
In[·]:= With[{h = 0.01, k = 0.01}, Sum[h*k, {x, 0, 1, h},
            {y, 0, f[x], k}]]
Out[·]:= 0.3444
```

```
In[·]:= N[Integrate[Integrate[1, {y, 0, f[x]}],
            {x, 0, 1}]]
Out[·]:= 0.333333
```

This is an example of an *iterated integral* which effectively computes the so-called *area integral*.

Now let us take a function $f : D \to \mathbb{R}$ defined over a closed rectangular area $D \subset \mathbb{R}^2$ and consider the problem that motivated the definition of the Riemann integral: how to compute the volume of the solid object enclosed between the xy-plane and the graph of f? We can imitate the method of *Darboux sums* described in Volume 1, that is, we partition the rectangle into smaller rectangles (e. g., by partitioning the sides of the rectangle into intervals of equal length). We can then form lower and upper Darboux sums as in the one variable case. If the supremum of lower sums is equal to the infimum of the uppers sums, where the supremum and infimum are taken over all partitions of the rectangle into smaller rectangles, then we say that the function is *Riemann integrable over the rectangle*. The illustration below shows a typical lower Darboux sum for the function $f(x,y) = x^2+y^2$ over the square $0 \le x, y \le 1$, that is, the total volume of cuboids on the rectangles of the partition (i. e., rectangular boxes) under the graph of the function. The function in this example is continuous but the integral will be defined also for a function for which the set $\{x \in D, f \text{ is not continuous at } x\}$ has measure 0 (where measure is defined as in [13]). The *Lebesgue integral* is defined in a more general way: it is also the supremum of volumes of cylinders lying below the graph but whose bases are disjoint "measurable sets" whose set-theoretic sum is the entire base rectangle. A rectangle is always a Lebesgue measurable set but Lebesgue measurable sets can be more complicated than rectangles. Nevertheless, it is easy to see that if a function is Riemann integrable, then its Riemann and Lebesgue integrals coincide.

```
In[·]:= Clear[f]; f[x_, y_] := x^2 + y^2
In[·]:= g1 = Plot3D[f[x, y], {x, 0, 1}, {y, 0, 1}];
In[·]:= g2 = With[{h = 0.1, k = 0.1},
        Show[Graphics3D[Table[{RGBColor[
        Random[], Random[], Random[]],
        Cuboid[{x, y, 0}, {x + h, y + k,
        f[x, y]}]}, {x, 0, 1 - h, h},
        {y, 0, 1 - k, k}]]]];
In[·]:= Show[g1, g2]
```

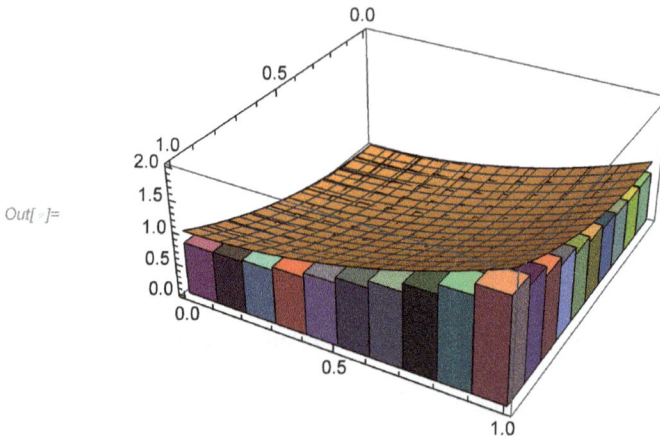

Out[·]=

Figure 5.3

5.2 Integrals over nonrectangular sets

Once we can integrate over rectangles, we can integrate over other measurable sets. The idea is that, to compute the integral of a function f over a set A, we choose a rectangle C (possibly infinite) containing A. We then extend the function f to a function $\tilde{f}$ on C, by defining it equal to f on A and to 0 on the complement $C \setminus A$. If f is actually already defined over $\mathbb{R}^2$, then we can define $\tilde{f}(x)$ as the product $f(x)\chi_A$, where χ_A is the characteristic function of A ($\chi_A(x) = 1$ if $x \in A$ and $\chi_A(x) = 0$ if $x \notin A$). In order for the Riemann integral to be defined in this way, χ_A has to be integrable, which is equivalent to the requirement that the boundary of A is a set of measure 0 (see [13]). For example, this will be the case if the boundary of A is a finite union of graphs of continuous functions (this is the definition of integrability used in [10, Theorem 2 of Section 5.2]). If the Lebesgue approach is used, the set A only needs to be Lebesgue measurable, which is a much more general condition. Here we shall only deal with cases which satisfy the definition of integrability as in [10]. This can be extended to general $\mathbb{R}^n$.

Let us see how this is implemented in Mathematica®. Suppose we want to integrate a function f over the unit disk, where f is given as above by $f(x,y) = x^2 + y^2$. If we think of f as defined only on the unit disk (in other words, we consider the restriction of f to the unit disc), we can define $\tilde{f}$ (which we will denote by g) using the function Piecewise:

```
In[·]:= g[x_, y_] := Piecewise[{{f[x, y], x^2 + y^2 <= 1},
         {0, x^2 + y^2 > 1}}]
```

```
In[·]:= gr = Plot3D[g[x, y], {x, -2, 2}, {y, -2, 2}, Boxed
         -> False, Axes -> False, PlotRange -> All]
```

Out[ ]=

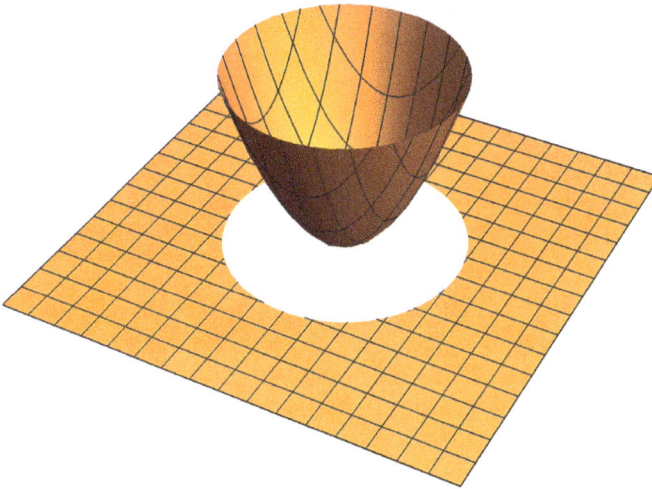

Figure 5.4

We can now integrate this g over the whole plane:

```
In[·]:= Integrate[g[x, y], {x, -Infinity, Infinity},
         {y, -Infinity, Infinity}]
```
$$Out[·]:= \frac{\pi}{2}$$

Of course, we could have equally well integrated over any rectangle containing the unit disc, e. g.,

```
In[·]:= Integrate[g[x, y], {x, -2, 2}, {y, -2, 2}]
```
$$Out[·]:= \frac{\pi}{2}$$

Alternatively, since *f* is actually defined on the whole plane, we can multiply it by the characteristic function of the unit disk and integrate this product. The characteristic function is most easily defined in Mathematica® by using the function Boole:

In[·]:= h[x_, y_] := Boole[x^2 + y^2 <= 1]*f[x, y]

In[·]:= Integrate[h[x, y], {x, -2, 2}, {y, -2, 2}]
Out[·]:= $\frac{\pi}{2}$

Mathematica® can compute some very difficult integrals exactly, often expressing the answer in terms of special functions. For example,

In[·]:= Integrate[Boole[x^4 + y^6 <= 1]*
    (x^2 + y^2), {x, -2, 2}, {y, -2, 2}]
Out[·]:= $-\frac{2\left(4\pi\,\text{Gamma}[7/4]^2 - \sqrt{\pi}\,\text{Gamma}[-1/6]\,\text{Gamma}[5/4]\,\text{Gamma}[23/12]\right)}{3\,\text{Gamma}[-1/6]\,\text{Gamma}[7/4]\,\text{Gamma}[23/12]}$

In more complicated cases, where no explicit formula can be obtained, one can still get a numerical answer (with arbitrary precision) by using NIntegrate:

In[·]:= NIntegrate[Boole[x^2 + y^2 <= 1]*
    Sin[Exp[Abs[x*y]]], {x, -2, 2},
    {y, -2, 2}, WorkingPrecision -> 10]
Out[·]:= 2.870966474

5.3 Fubini' s theorem

Fubini' s theorem makes it possible to convert the problem of computing an integral over a subset of $\mathbb{R}^n$ into one of computing *n* iterated integrals of one variable. We will only quote here the statement given in [14, Theorem 8.5.1] in the case of Lebesgue integrals. The statement for Riemann integrals can be found in [13, 3–10]. We will follow [14] and limit ourselves only to the statement in the case *n* = 2, because while its extension to the general case is obvious, it is rather cumbersome to write down.

Theorem 10 ([14, Theorem 8.5.1] Fubini' s theorem). *Let* $f : \mathbb{R}^2 \to \mathbb{R}$ *be an absolutely integrable function. Then there exist absolutely integrable functions* $F : \mathbb{R} \to \mathbb{R}$ *and* $G : \mathbb{R} \to \mathbb{R}$ *such that for almost every* $x, f(x,y)$ *is absolutely integrable in y with*

$$F(x) = \int_{\mathbb{R}} f(x,y)\,dy$$

and for almost every $y, f(x,y)$ *is absolutely integrable in x with*

$$G(y) = \int_{\mathbb{R}} f(x,y)\,dx.$$

Finally, we have

$$\iint_{\mathbb{R}^2} f(x,y)\, dx\, dy = \int_{\mathbb{R}} F(x)\, dx = \int_{\mathbb{R}} G(y)\, dy.$$

This theorem allows us to compute two-dimensional integrals by splitting them into one-dimensional ones. It is important to remember that the functions $F(x)$ and $G(y)$ are defined for almost all x and y, where "almost all" means that the sets of points where F and G are not defined have measure zero. For example, consider the function on $\mathbb{R}^2$ given by

In[·]:= f[x_, y_] := Piecewise[{{1, y > 0 && x == 0},
         {-1, y < 0 && x == 0}}]

The integral of this function over $\mathbb{R}^2$ is, of course, 0:

In[·]:= Integrate[f[x, y], {x, -Infinity, Infinity},
         {y, -Infinity, Infinity}]
Out[·]:= 0

However, $F(0)$ does not exist:

In[·]:= Integrate[f[0, y], {y, -Infinity, Infinity}]

$$Out[\cdot]:= \int_{-\infty}^{\infty} \left(\begin{array}{ll} 1 & y > 0 \\ -1 & y < 0 \\ 0 & \text{True} \end{array} \right) dy$$

Sometimes the following notation is used: $dA = dx\, dy$ (dA is then called the *area element*). Hence, $\iint_A f(x,y)\, dA$ is called an *area integral*. Informally, we can think of it as the area formed by infinitesimal rectangles of the kind that appeared at the beginning of the chapter.

5.3.1 Example 1

Let us calculate the volume under the surface $f(x,y) = x^2 + y^2$ and above the rectangle $0 \le x, y \le 1$.

In[·]:= Integrate[x^2 + y^2, {x, 0, 1}, {y, 0, 1}]
Out[·]:= $\dfrac{2}{3}$

In[·]:= Integrate[x^2 + y^2, {y, 0, 1}, {x, 0, 1}]
Out[·]:= $\dfrac{2}{3}$

In fact, the volume under the surface can also be computed using a simple *triple integral* (or a *volume integral*) by integrating the function 1:

In[·]:= `Integrate[Integrate[1, {z, 0, x^2 + y^2}],`
`{y, 0, 1}, {x, 0, 1}]`

Out[·]:= $\dfrac{2}{3}$

5.3.2 Example 2

Compute the area of a disc of radius *r*.

We can take the disk with center at $(0,0)$ and radius *r*, given by $x^2 + y^2 \leq r^2$. We know that we can compute the area by using the method of Volume 1, i. e.,

In[·]:= `2*Integrate[Sqrt[r^2 - x^2], {x, -r, r},`
`Assumptions -> r > 0]`

Out[·]:= πr^2

But we can obtain it by integrating the function 1 over the disk $x^2 + y^2 \leq r^2$, i. e.,

In[·]:= `Integrate[Boole[x^2 + y^2 <= r^2],`
`{x, -Infinity, Infinity}, {y, -Infinity,`
`Infinity}, Assumptions -> r > 0]`

Out[·]:= πr^2

By Fubini's theorem, this computation can be performed as follows in two ways: either by dividing the disk into "vertical strips", then integrating along them, and then integrating the resulting function "horizontally", or by performing vertical integration first and then horizontal one. The first case is illustrated below. We first think of *x* as an independent variable and of *y* as a variable that depends on *x*. This is the same as "dividing the disc by vertical sections".

In[·]:= `gr1 = Graphics[Circle[{0, 0}, 1],`
`Axes -> True, AspectRatio -> Automatic,`
`AxesLabel -> {"x axis", "y axis"}];`

In[·]:= `vs[a_] := Graphics[{Red, Line[{{a,`
`-Sqrt[1 - a^2]}, {a, Sqrt[1 - a^2]}}]}];`

In[·]:= `txt[a_] := Graphics[{Text[StyleForm[`
`TraditionalForm[y == Sqrt[1 - a^2]]],`
`{a, (10/9)*Sqrt[1 - a^2]}], Text[`
`StyleForm[TraditionalForm[{a, 0}]],`
`{a, 0.05}], Text[StyleForm[`
`TraditionalForm[y == -Sqrt[1 - a^2]]],`
`{a, (-10/9)*Sqrt[1 - a^2]}]}];`

In[·]:= `Manipulate[Show[gr1, vs[a], txt[a]],`
`{{a, 0.3, "a"}, -1, 1},`
`SaveDefintions -> True]`

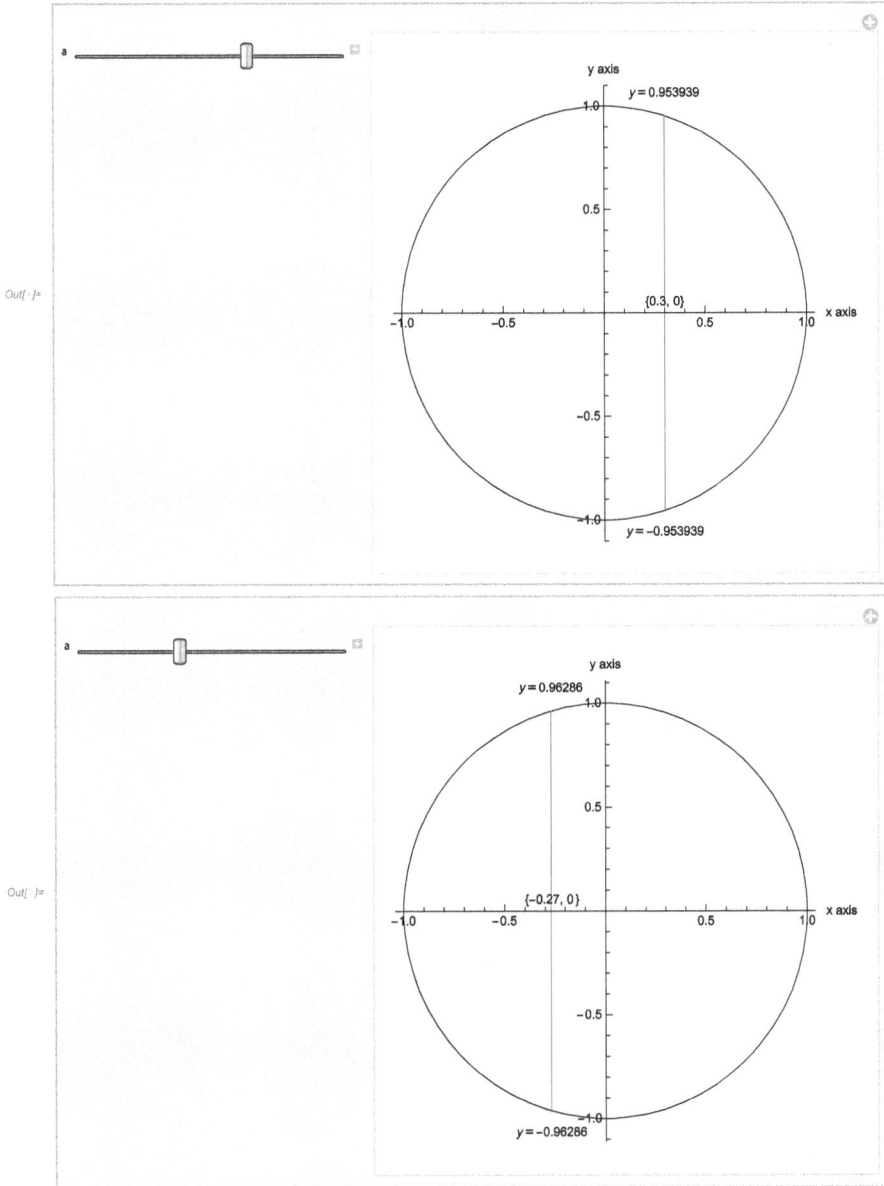

Figure 5.5

The disc is now described by the inequalities $-r \le x \le r, -\sqrt{r^2 - x^2} \le y \le \sqrt{r^2 - x^2}$. We first compute

```
In[·]:= Integrate[1, {y, -Sqrt[r^2 - x^2],
        Sqrt[r^2 - x^2]}]
```

$$Out[\cdot]:= 2\sqrt{r^2 - x^2}$$

and then

> *In[·]:=* `Integrate[2*Sqrt[r^2 - x^2], {x,`
> `-r, r}, Assumptions -> r > 0]`
> *Out[·]:=* $\pi\, r^2$

The second approach is completely analogous, since everything is symmetrical with respect to the variables x and y.

5.3.3 Example 3

Compute the integral of the function $f : \mathbb{R}^2 \to \mathbb{R}$ given by $f(x, y) = x + 2y$ over the set enclosed between the parabola $y = x^2$ and the straight line $y = x + 2$.

The region we are integrating over can be visualized using the function `Region-Plot`. First, we can find the two "corners" as follows:

> *In[·]:=* `{x, x + 2} /. Solve[x^2 == x + 2, x]`
> *Out[·]:=* `{{-1, 1}, {2, 4}}`

> *In[·]:=* `RegionPlot[x^2 <= y <= x + 2,`
> `{x, -1, 2}, {y, 0, 4}, Axes`
> `-> True, Frame -> False]`

Out[·]=

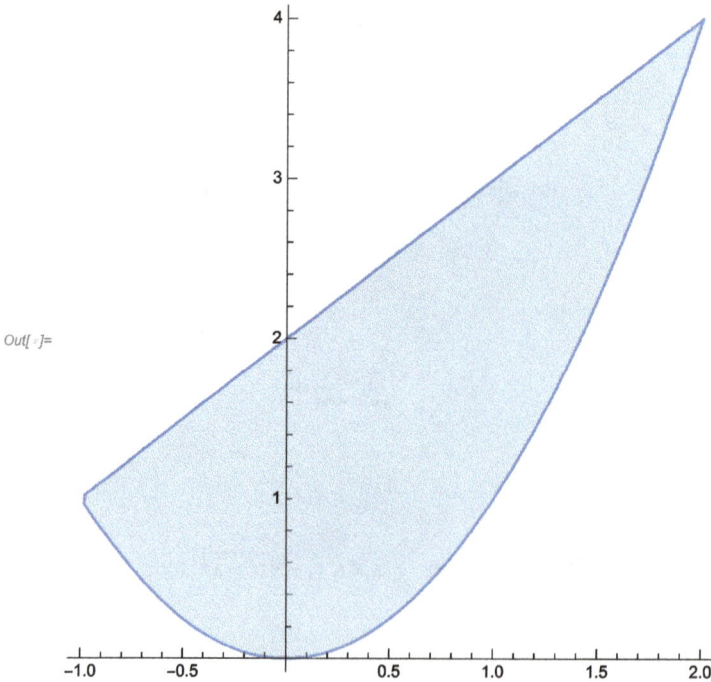

Figure 5.6

Mathematica® can compute the answer itself by

In[·]:= `Integrate[(x + 2*y)*Boole[x^2 <=`
            `y <= x + 2], {x, -Infinity, Infinity},`
            `{y, -Infinity, Infinity}]`
Out[·]:= $\dfrac{333}{20}$

We can also do it by applying Fubini's theorem in two different ways. The first way is to choose x as the independent variable and describe the region in the form $x^2 \le y \le x + 2$, $-1 \le x \le 2$, and compute

In[·]:= `Integrate[x + 2*y, {x, -1, 2}, {y, x^2, x + 2}]`
Out[·]:= $\dfrac{333}{20}$

We can also describe the area as the union of two sets: $0 \le y \le 1$, $-\sqrt{y} \le x \le \sqrt{y}$ and $1 \le y \le 4$, $y - 2 \le x \le \sqrt{y}$. Thus the integral is also

In[·]:= `Integrate[Integrate[x + 2*y, {x,`
            `-Sqrt[y], Sqrt[y]}], {y, 0, 1}] +`
            `Integrate[Integrate[x + 2*y, {x,`
            `y - 2, Sqrt[y]}], {y, 1, 4}]`
Out[·]:= $\dfrac{333}{20}$

5.4 Change of variables theorem

The next theorem often makes it possible to compute integrals of functions of many variables by a method which generalizes the change of variable formula from the theory of integration of functions of one variable. We state it without proof (a proof in a special case for the Riemann integral can be found in [10, Section 6.2]).

First, recall that a *diffeomorphism* $\phi : D_1 \to D_2$, where D_1 and D_2 are subsets of some Euclidean space $\mathbb{R}^n$, is a differentiable bijective mapping whose inverse $\phi^{(-1)} : D_2 \to D_1$ is also differentiable.

Theorem 11. *Let $\Omega \subset \mathbb{R}^n$ be an open set and let $\phi : \Omega \to \phi(\Omega)$ be a diffeomorphism of class C^1 of Ω onto $\phi(\Omega)$. Let f be an integrable function on $\phi(\Omega)$. Then $f \circ \phi$ is integrable and*

$$\int_{\phi(\Omega)} f(x) \, d\lambda = \int_{\Omega} \|D\phi\| f(\phi(x)) \, d\lambda,$$

where $d\lambda$ refers to the generalized volume element.

5.4.1 Example 4

Compute the integral of the function $f(x, y) = x$ over the parallelogram with vertices at $(0, 0)$, $(2, 1)$, $(1, 1)$, and $(3, 2)$.

```
In[·]:= Graphics[{Green, Polygon[{{0, 0},
        {1, 1}, {3, 2}, {2, 1}}]}]
```

Out[ ]=

Figure 5.7

We can, of course, solve this problem using Fubini's theorem, but it is slightly easier to use a coordinate change to transform the integral into one over a square. We can easily find a transformation that takes the unit square with one vertex at the origin into the parallelogram. The matrix of this transformation is given by

```
In[·]:= M = Transpose[{{1, 1}, {2, 1}}]
Out[·]:= {{1, 2}, {1, 1}}
```

```
In[·]:= % // MatrixForm
```
$$Out[·]//MatrixrForm=\begin{pmatrix} 1 & 2 \\ 1 & 1 \end{pmatrix}$$

So

```
In[·]:= phi[x_, y_] := M . {x, y};
```

```
In[·]:= {phi[0, 0], phi[0, 1], phi[1, 1], phi[1, 0]}
Out[·]:= {{0, 0}, {2, 1}, {3, 2}, {1, 1}}
```

```
In[·]:= f[{x_, y_}] := x
```

```
In[·]:= Graphics[{Green, Polygon[{{0, 0},
        {1, 1}, {3, 2}, {2, 1}}], Opacity[0.5],
        Red, Polygon[{{0, 0}, {0, 1}, {1, 1},
        {1, 0}}]}], Axes -> True]
```

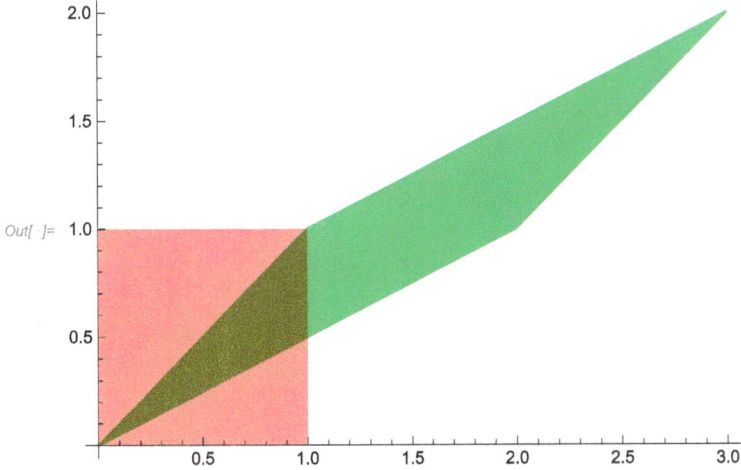

Figure 5.8

Hence, we only need to compute

```
In[·]:= Integrate[Abs[Det[M]]*f[phi[x, y]],
        {x, 0, 1}, {y, 0, 1}]
```

$$Out[\cdot]:= \frac{3}{2}$$

We can confirm the answer integrating over the original polygon:

```
In[·]:= Integrate[x*Boole[y <= x && y >= x/2 &&
        y <= (x + 1)/2 && y >= x - 1],
        {x, 0, Infinity}, {y, 0, Infinity}]
```

$$Out[\cdot]:= \frac{3}{2}$$

5.4.2 Example 5: polar coordinates

Compute the integral of the function $f(x, y) = x^2 + y^2$ over the annulus of inner radius 1 and outer radius 2.

```
In[·]:= RegionPlot[1 <= x^2 + y^2 <= 2,
        {x, -2, 2}, {y, -2, 2}]
```

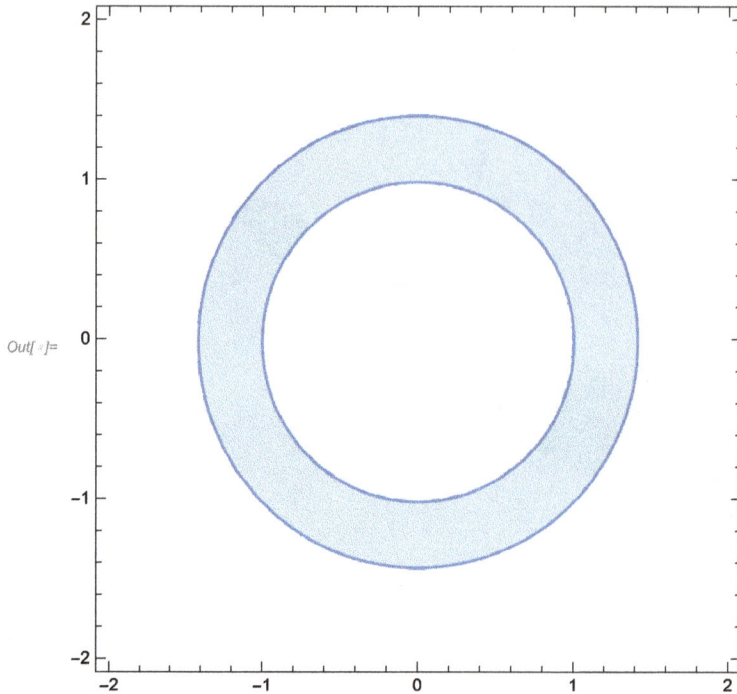

Figure 5.9

To solve this kind of problem, it is convenient to change to the polar coordinate system. The *polar coordinate system* can be viewed as a transformation from one plane (on the left in the illustration below) with the Cartesian coordinates (r, t) into another system with the Cartesian coordinates (x, y), where the transformation is given by $(r, t) \rightarrow (r \cos t, t \sin t)$. In the illustration below, a point (vector) in the first plane is represented by a red arrow, its image in the plane on the right by a blue arrow. Of course, the transformation is not a diffeomorphism, since vertical lines are taken into circles. However, if we restrict the transformation to the open set $(0, +\infty) \times (0, 2\pi)$, it becomes a diffeomorphism onto its image which is $\mathbb{R}^2 \setminus \mathbb{R}_+$, where $\mathbb{R}_+$ is the nonnegative real axis. Because the measure of $\mathbb{R}_+$ is 0 (in $\mathbb{R}^2$), and the integral over a set of measure 0 is 0, the change of variable theorem still holds.

```
In[·]:= Manipulate[GraphicsGrid[{{Graphics[
          {PointSize[0.02], Red, Arrow[{{0, 0},
          {r, t}}]]}, Axes -> True, AxesLabel ->
          {"r", "t"}, PlotRange -> {{0, 1},
          {0, 2*Pi}}], Graphics[{PointSize[
```

```
0.04], Blue, Arrow[{{0, 0},
{r*Cos[t], r*Sin[t]}}]]}, Axes
-> True, AxesLabel -> {"x", "y"},
PlotRange -> {{-1, 1}, {-1, 1}}]]}],
{{r, 0.2, "r"}, 0, 1}, {{t,
Pi/3, "t"}, 0, 2*Pi}]
```

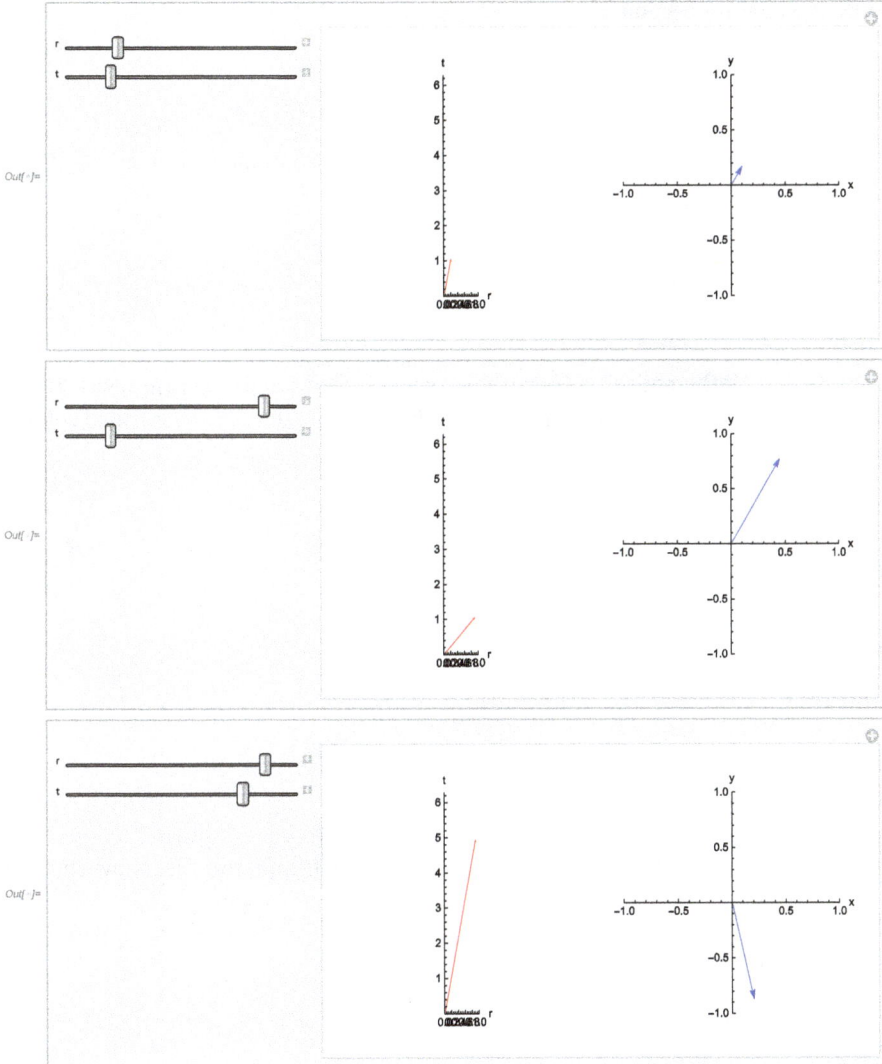

Figure 5.10

The function f can be now viewed as $(r, t) \rightarrow r^2(= x^2 + y^2)$. We now compute the absolute value of the Jacobian determinant:

In[·]:= Assuming[r > 0, Simplify[Abs[Det[
          D[{r*Cos[t], r*Sin[t]}, {{t, r}}]]]]]
Out[·]:= r

Hence the integral becomes

In[·]:= Integrate[r^2*r, {r, 1, 2}, {t, 0, 2*Pi}]
Out[·]:= $\dfrac{15\pi}{2}$

We can check the answer by integrating in the original coordinate system:

In[·]:= Integrate[(x^2 + y^2)*Boole[1 <=
          x^2 + y^2 <= 4], {x, -Infinity,
          Infinity}, {y, -Infinity, Infinity}]
Out[·]:= $\dfrac{15\pi}{2}$

Note that **Mathematica**®'s CoordinateTransformData and CoordinateChartData will give us the required transformation and the value of $\|D\phi\|$, which is called *volume factor*:

In[·]:= CoordinateTransformData["Polar" ->
          "Cartesian", "Mapping", {r, t}]
Out[·]:= {r Cos[t], r Sin[t]}

In[·]:= CoordinateChartData["Polar",
          "VolumeFactor", {r, t}]
Out[·]:= r

5.4.3 Example 6: spherical coordinates

Compute the integral of the function $f(x, y, z) = xy$ over the quarter-disc shown below:

In[·]:= RegionPlot3D[x > 0 && y > 0 && z > 0
          && x^2 + y^2 + z^2 <= 1, {x, -1, 1},
          {y, -1, 1}, {z, -1, 1}, Axes -> False,
          AspectRatio -> 1]

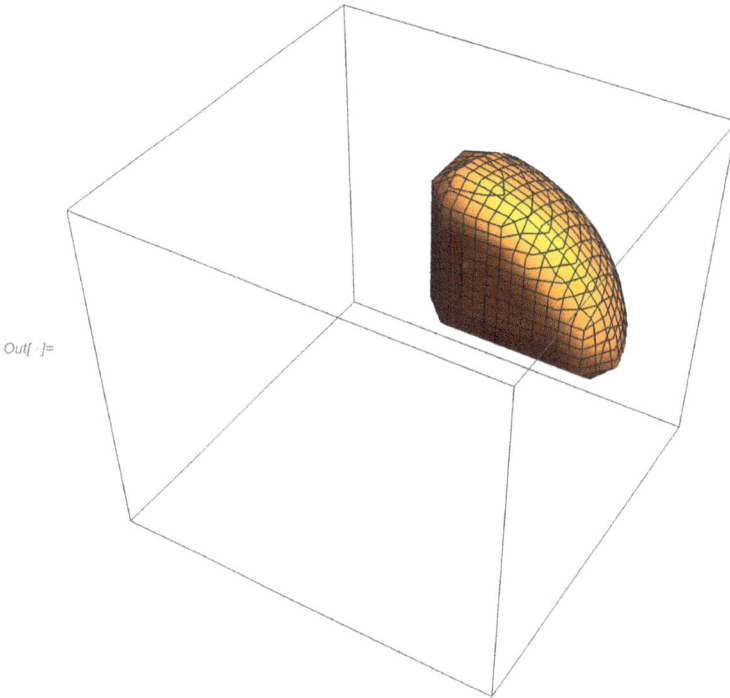

Out[ ]=

Figure 5.11

Using Mathematica® makes this very easy:

In[·]:= Integrate[x*y*Boole[x > 0 && y > 0
&& z > 0 && x^2 + y^2 + z^2 <= 1],
{x, -Infinity, Infinity}, {y,
-Infinity, Infinity}, {z,
-Infinity, Infinity}]

Out[·]:= $\dfrac{1}{15}$

If we had to do it manually, it would be preferable to use a change of coordinates rather than Fubini's theorem. In this case the right choice is to use the *spherical coordinates* given by

In[·]:= CoordinateTransformData["Spherical"
-> "Cartesian", "Mapping", {r, u, v}]

Out[·]:= {r Cos[v] Sin[u], r Sin[u] Sin[v], r Cos[u]}

The determinant of the Jacobian is

In[·]:= FullSimplify[Det[D[%, {{r, u, v}}]]]

Out[·]:= r^2 Sin[u]

which can also be obtained by using CoordinateChartData

In[·]:= CoordinateChartData["Spherical",
        "VolumeFactor", {r, u, v}]
Out[·]:= r^2 Sin[u]

Hence we get

In[·]:= Integrate[r*Cos[v]*Sin[u]*r*Sin[u]*
        Sin[v]*r^2*Sin[u], {u, 0, Pi/2},
        {v, 0, Pi/2}, {r, 0, 1}]
Out[·]:= $\dfrac{1}{15}$

This again agrees with the answer obtained using Boole.

5.4.4 Example 7: cylindrical coordinates

Compute the integral of the function $f(x, y, z) = |xyz|$ over the region $S = \{(x, y, z) \in \mathbb{R}^3, 1 \geq z \geq 3\sqrt{x^2 + y^2}\}$.

The region can be visualized using RegionPlot3D:

In[·]:= RegionPlot3D[1 >= z >= 3*Sqrt[x^2
        + y^2], {x, -1, 1}, {y, -1, 1},
        {z, 0, 1}]

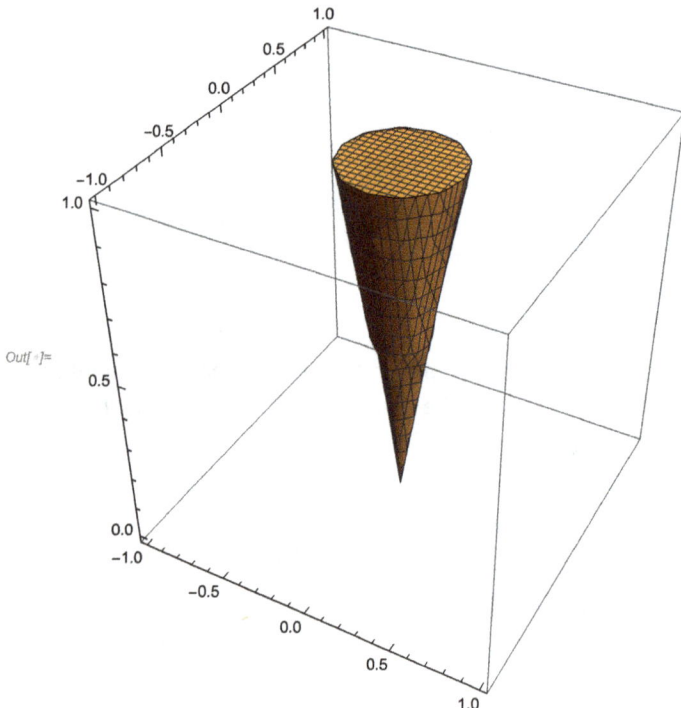

Out[·]=

Figure 5.12

Mathematica® can, of course, compute the integral directly:

In[·]:= `Integrate[Abs[x*y*z]*Boole[1 >=`
`z >= 3*Sqrt[x^2 + y^2]], {x,`
`-Infinity, Infinity}, {y,`
`-Infinity, Infinity}, {z,`
`-Infinity, Infinity}]`

Out[·]:= $\dfrac{1}{972}$

For computing the integral by hand, it is most convenient to use the change of variable theorem and switch to *cylindrical coordinates.* They are given by

In[·]:= `CoordinateTransformData["Cylindrical"`
`-> "Cartesian", "Mapping", {r, t, z}]`

Out[·]:= `{r Cos[t], r Sin[t], z}`

The volume factor is

In[·]:= `CoordinateChartData["Cylindrical",`
`"VolumeFactor", {r, t, z}]`

Out[·]:= `r`

Hence the answer is

In[·]:= `Integrate[r*Abs[r*Cos[t]*r*Sin[t]*z],`
`{z, 0, 1}, {r, 0, z/3}, {t, 0, 2*Pi}]`

Out[·]:= $\dfrac{1}{972}$

5.4.5 Example 8

Compute the volume of the solid torus obtained by rotating a disc of radius *r* and center at $(R, 0, 0)$ (where $0 < r < R$) in the *xz*-plane about the *z* axis.

We first use Mathematica®'s RotationTransform to obtain a mapping that rotates a circle in the required way through a given angle *v*:

In[·]:= `RotationTransform[v, {0, 0, 1}][`
`{R, 0, 0} + {r*Cos[u], 0, r*Sin[u]}]`

Out[·]:= `{(R + r Cos[u]) Cos[v], (R + r Cos[u]) Sin[v], r Sin[u]}`

To make an illustration, we need to choose concrete values for *r* and *R*. We take $r = 1$ and $R = 2$:

In[·]:= `ParametricPlot3D[% /. {r -> 1, R -> 2},`
`{u, 0, 2*Pi}, {v, 0, 2*Pi}, Axes -> False]`

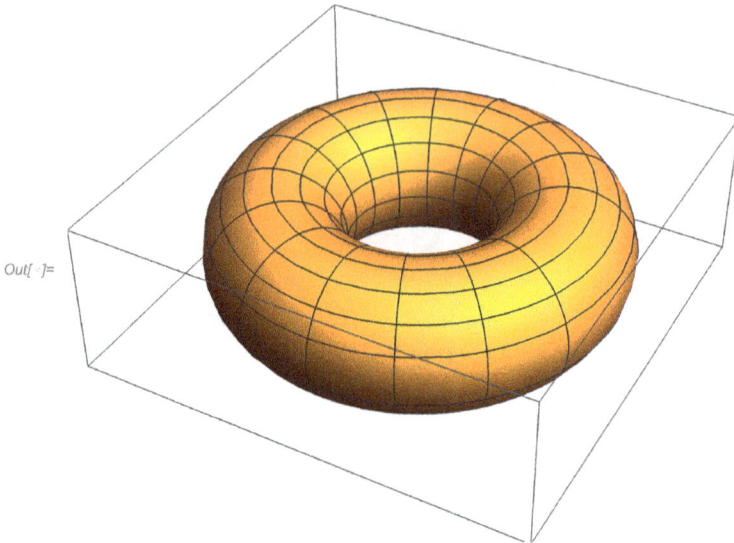

Out[ ]=

Figure 5.13

We can again view the parametric equations as a diffeomorphism from the set $(0, R) \times (0, 2\pi) \times (0, 2\pi)$ onto its image. The "missing part" of the torus over which integration is performed has measure 0 (in $\mathbb{R}^3$), so it does not affect the value of the integral. The volume factor is

In[·]:= `Simplify[Abs[Det[D[{Cos[v]*(r*Cos[u] +`
`        R), Sin[v]*(r*Cos[u] + R), r*Sin[u]},`
`        {{r, u, v}}]]]]`
Out[·]:= `Abs[r (R + r Cos[u])]`

The integral of the function 1 over the torus, which gives its volume, is

In[·]:= `Integrate[Abs[s*(s*Cos[u] + R)],`
`        {s, 0, r}, {u, 0, 2*Pi}, {v, 0, 2*Pi},`
`        Assumptions -> R > s > 0]`
Out[·]:= $2\pi^2 r^2 R$

6 Stokes' theorem

There are three main integral theorems of vector analysis:

Green's theorem:

$$\int_{\partial D} (P\,dx + Q\,dy) = \iint_D \left(\frac{\partial Q}{\partial x} - \frac{\partial P}{\partial y} \right) dx\,dy;$$

Stokes' theorem:

$$\int_{\partial S} \mathbf{F} \cdot d\mathbf{s} = \iint_S (\nabla \times \mathbf{F}) \cdot d\mathbf{S} = \iint_S \text{curl}\,\mathbf{F} \cdot d\mathbf{S};$$

Gauss' (Ostrogradsky's, divergence) theorem:

$$\iint_{\partial W} \mathbf{F} \cdot d\mathbf{S} = \iiint_W (\nabla \cdot \mathbf{F})\,dV.$$

In this section we first discuss Green's theorem and then Stokes' theorem and its proof. We also give a few applications of Stokes' theorem. We use [10] as the main source for definitions and statements of theorems. We also use [2, 9] for theory and examples and [16] for physical interpretation. Stokes' theorem has numerous applications in electromagnetism and fluid dynamics. It can also be written using the language of differential forms.

Stokes' theorem is also known as *Kelvin–Stokes' theorem* or the *curl theorem*. It was discovered around 1850. The history of Stokes' theorem is very complicated. We refer the interested reader to [6].

6.1 Vector fields

A *vector field* on $\mathbb{R}^n$ is a map $\mathbf{F} : A \subset \mathbb{R}^n \to \mathbb{R}^n$ that assigns a vector to each point. A *flow* for $\mathbf{F}$ is a path satisfying $\mathbf{c}'(t) = \mathbf{F}(\mathbf{c}(t))$. Let us visualize a moving particle for the 2D vector field $(-y, x)$:

```
In[·]:= DSolve[{Derivative[1][x][t] == -y[t],
        Derivative[1][y][t] == x[t]}, {x[t], y[t]}, t]
Out[·]:= {{x[t] → c₁ Cos[t] − c₂ Sin[t], y[t] → c₂ Cos[t] + c₁ Sin[t]}}
```

For $c_1 = c_2 = 1$, we have

```
In[·]:= Manipulate[Show[{Graphics[{PointSize[Large], Red,
        Evaluate[Point[{Cos[t] - Sin[t], Cos[t] + Sin[t]}]]}],
        Graphics[VectorPlot[{-y, x}, {x, -3, 3},
        {y, -3, 3}]]}], {t, 0.1, 30}]
```

https://doi.org/10.1515/9783110660395-006

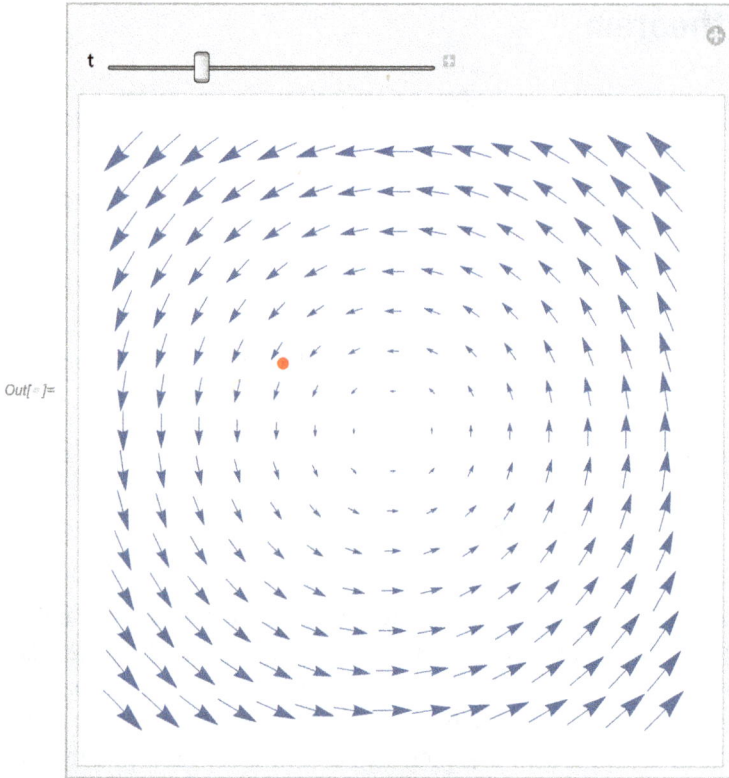

Figure 6.1

A vector field in $\mathbb{R}^3$ has the form $\mathbf{F}(x,y,z) = (F_1(x,y,z),\, F_2(x,y,z),\, F_3(x,y,z))$. One physical interpretation is, for instance, the velocity field of a fluid. The *curl of the vector field* $\mathbf{F}$ is a new vector field defined by

$$\text{curl } \mathbf{F} = \nabla \times \mathbf{F} = \left(\frac{\partial F_3}{\partial y} - \frac{\partial F_2}{\partial z},\, \frac{\partial F_1}{\partial z} - \frac{\partial F_3}{\partial x},\, \frac{\partial F_2}{\partial x} - \frac{\partial F_1}{\partial y} \right).$$

For vector fields of the form $\mathbf{F} = (P(x,y),\, Q(x,y),\, 0)$ there is a simple expression for curl $\mathbf{F}$:

$$\text{curl } \mathbf{F} = \nabla \times \mathbf{F} = \left(0,\, 0,\, \frac{\partial Q}{\partial x} - \frac{\partial P}{\partial y} \right).$$

A natural physical interpretation is the following: the curl of a vector field $\mathbf{F}$ describes the rotary motions caused by it. The direction of the curl is along the axis of rotation and the magnitude is the magnitude of rotation. For a 3D vector field $(-y, x, 0)$, the curl in 3D is visualized as follows:

```
In[·]:= Curl[{-y, x, 0}, {x, y, z}]
Out[·]:= {0, 0, 2}
```

In[·]:= VectorPlot3D[%, {x, -1, 1}, {y, -1, 1}, {z, -1, 1}]

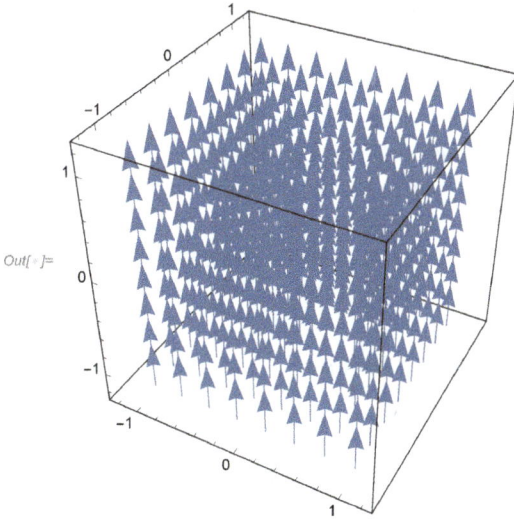

Out[·]=

Figure 6.2

We have rotation around z-axis.

Other operators, important in vector analysis, are the *gradient vector field* (given by $\nabla f(x, y, z) = (\frac{\partial f}{\partial x}, \frac{\partial f}{\partial y}, \frac{\partial f}{\partial z})$) and the *divergence of a vector field* **F**, which is a scalar field div **F** $= \nabla \cdot \mathbf{F} = \frac{\partial F_1}{\partial x} + \frac{\partial F_2}{\partial y} + \frac{\partial F_3}{\partial z}$. A natural physical interpretation is that divergence measures the rate of expansion per unit volume under the flow of a fluid. These operators are related by a number of useful identities: the curl of a gradient is zero for any C^2 function f, i. e., $\nabla \times (\nabla f) = 0$:

In[·]:= Curl[Grad[f[x, y, z], {x, y, z}], {x, y, z}]
Out[·]:= {0, 0, 0}

The divergence of a curl is zero for any C^2 vector field **F**, i. e., div curl **F** $= \nabla \cdot (\nabla \times \mathbf{F}) = 0$:

In[·]:= Div[Curl[{F1[x, y, z], F2[x, y, z],
        F3[x, y, z]}, {x, y, z}], {x, y, z}]
Out[·]:= 0

6.2 Integration

6.2.1 Line integrals (integration of a vector field along a path)

For a vector field **F** continuous on a (piecewise) C^1 path **c** : $[a, b] \to \mathbb{R}^3$, the *line integral* is defined by

$$\int_{\mathbf{c}} \mathbf{F} \cdot d\mathbf{s} = \int_a^b \mathbf{F}(\mathbf{c}(t)) \cdot \mathbf{c}'(t) \, dt.$$

In this definition the choice of orientation matters. The physical interpretation of such line integrals is the work done by a force field **F** during the motion along the path **c**. It is also the circulation of the velocity field of a fluid.

A *conservative field* is defined as the gradient of a potential. Let Ω in $\mathbb{R}^2$ or $\mathbb{R}^3$ be an open, simply connected region. Consider a vector field of class C^1 on Ω. Then the vector field is *conservative* if and only if it is curl-free (i. e., its curl is zero). A *simply connected set* is a set for which any closed loop in Ω can be continuously deformed to a point, staying always in Ω (e. g., convex sets, star-shaped sets). For the continuous conservative vector field $\mathbf{F} = \nabla f$, the line integral over the piecewise regular curve $\mathbf{c} : [a, b] \to \mathbb{R}^3$ can easily be calculated:

$$\int_c \mathbf{F} \cdot d\mathbf{s} = f(\mathbf{c}(b)) - f(\mathbf{c}(a)).$$

6.2.2 Surface integrals of vector fields (flux integrals)

Let us assume that the surface S is oriented by a unit normal vector field **n**. The *surface integral of the vector field* **F** is defined as the integral of the normal component of the vector field over the surface:

$$\iint_S \mathbf{F} \cdot d\mathbf{S} = \iint_S \mathbf{F} \cdot \mathbf{n} \, dS.$$

In this definition the choice of orientation matters. The usual physical interpretation of surface integrals is as the rate of flow of a fluid, that is, the rate at which the fluid is crossing a given surface (the net quantity of fluid that flows across the surface per unit time).

In practice, surface integrals are calculated as follows. If a surface S is given by $z = g(x, y)$, $(x, y) \in D$, and is oriented by the upward-pointing normal, then we can calculate the surface integral by using the formula:

$$\iint_S \mathbf{F} \cdot d\mathbf{S} = \iint_D \left(F_1 \left(-\frac{\partial g}{\partial x} \right) + F_2 \left(-\frac{\partial g}{\partial y} \right) + F_3 \right) dx \, dy.$$

If the surface is the image $S = \Psi(D)$ under a function Ψ, where $\Psi : D \subset \mathbb{R}^2 \to \mathbb{R}^3$ is given by $\Psi(u, v) = (x(u, v), y(u, v), z(u, v))$, $(u, v) \in D$, then the surface integral of the vector field **F** defined on S is given by

$$\iint_S \mathbf{F} \cdot d\mathbf{S} = \iint_D \mathbf{F} \cdot (\mathbf{T}_u \times \mathbf{T}_v) \, du \, dv.$$

Here $\mathbf{T}_u = \frac{\partial}{\partial u}(x(u, v), y(u, v), z(u, v))$, $\mathbf{T}_v = \frac{\partial}{\partial v}(x(u, v), y(u, v), z(u, v))$.

6.3 Green's theorem

A *simple closed curve* is the image of a piecewise C^1 map $\mathbf{c} : [a, b] \to \mathbb{R}^2$ that is one-to-one on $[a, b)$ (no self-intersections) and $\mathbf{c}(a) = \mathbf{c}(b)$. A *simple region* D in $\mathbb{R}^2$ is given by $x \in [a, b]$, $\phi_1(x) \leq y \leq \phi_2(x)$ or by $y \in [c, d]$, $\psi_1(y) \leq x \leq \psi_2(y)$.

Theorem 12 (Green's theorem). *Let D be a simple region and let $\mathbf{c}$ be its boundary (a simple closed curve in $\mathbb{R}^2$). Suppose that the functions P, Q : D $\to$ $\mathbb{R}$ are of class C^1. Then the line integral can be related to the double integral over the region D enclosed by $\mathbf{c}$ as follows:*

$$\int_{\mathbf{c}} (P\,dx + Q\,dy) = \iint_D \left(\frac{\partial Q}{\partial x} - \frac{\partial P}{\partial y} \right) dx\,dy.$$

Orientation of the boundary $\mathbf{c}$ is chosen such that the region D is on the left as one moves along the boundary.

Note that for more complicated regions, one needs to divide them into elementary pieces.

Green's theorem is useful to calculate the area:

$$A = \frac{1}{2} \int_{\mathbf{c}} (x\,dy - y\,dx).$$

There also exist other forms of Green's theorem for the C^1 vector field $\mathbf{F} = (P(x, y), Q(x, y))$:

$$\int_{\partial D} \mathbf{F} \cdot d\mathbf{s} = \iint_D (\text{curl}\,\mathbf{F} \cdot \mathbf{k})\,dA,$$

$$\int_{\partial D} \mathbf{F} \cdot \mathbf{n}\,ds = \iint_D \text{div}\,\mathbf{F}\,dA,$$

where $\mathbf{n}$ is the outward unit normal to ∂D and $\mathbf{k}$ is the standard unit vector in $\mathbb{R}^3$.

6.4 Stokes' theorem

Theorem 13 (Stokes' theorem). *The line integral of the tangential component of the vector field $\mathbf{F}$ around the boundary ∂S is equal to the integral of the normal component of the curl of a vector field $\mathbf{F}$ over a surface S. That is, the line integral of a vector field around a (simple) closed curve is replaced by the flux integral of the curl of the vector field over any surface bounded by the curve.*

Green' s theorem is used in the proof of Stokes' theorem (the integrals on both sides of Stokes' formula are expressed in terms of double integrals over 2D domain).

In Stokes' theorem, orientation of the boundary curve is important (since the integrals on both sides depend on the orientation of the boundary curve and of the surface).

There are many assumptions for Stokes' theorem (see [10] for more details):

1. A compact, regular (a normal exists in each point) surface S, oriented by a smooth unit normal vector field **n**;
2. The surface S is bounded by a positively oriented (which means that the surface is on the left with the normal in the upright direction) simple (no self-intersections), closed, smooth curve C;
3. The surface S is given by $z = f(x,y)$, $(x,y) \in D$, where f is of class C^2, or by $\mathbf{r}(u,v) = (x(u,v), y(u,v), z(u,v))$, $(u,v) \in D$, where x, y, z are of class C^2 (we will need the equality of second order mixed partial derivatives in the proof of Stokes' theorem);
4. D is an open, bounded, simple region in $\mathbb{R}^2$ with a simple, closed, smooth boundary ∂D;
5. The boundary of D, ∂D in $\mathbb{R}^2$, is positively oriented (which means that D is on the left) and Green's theorem can be applied;
6. The boundary ∂D of the region D corresponds to the boundary curve C: if ∂D is parametrized by $(x(t), y(t))$ or by $(u(t), v(t))$, then the curve C is parametrized by $(x(t), y(t), f(x(t), y(t)))$ or by $(x(u(t), v(t)), y(u(t), v(t)), z(u(t), v(t)))$;
7. The vector field **F** on S is continuously-differentiable and continuous on the boundary C.

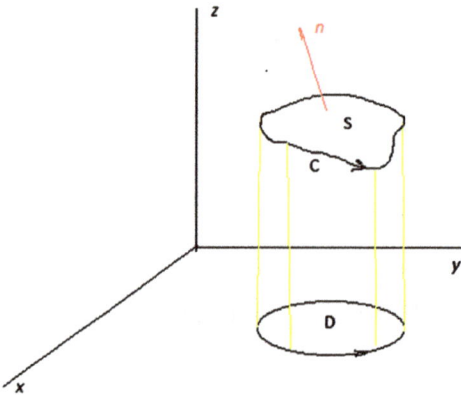

Figure 6.3

In[·]:= Show[{ParametricPlot3D[{r*Cos[u], r*Sin[u],
    Cos[r^2] + 2}, {r, 0, 1}, {u, 0, 2*Pi},
    Axes -> False, Boxed -> False, PlotStyle ->
    LightYellow, PlotRange -> {{-2, 2}, {-2, 2},
    {0, 3}}], ParametricPlot3D[{Cos[u], Sin[u],

```
Cos[1] + 2}, {u, 0, 2*Pi}, Axes -> False,
Boxed -> False, PlotStyle -> {Thick, Red}],
ParametricPlot3D[{r*Cos[u], r*Sin[u], 0},
{r, 0, 1}, {u, 0, 2*Pi}, Axes -> False,
Boxed -> False, PlotStyle -> LightYellow,
PlotRange -> {{-2, 2}, {-2, 2}, {0, 3}}],
ParametricPlot3D[{Cos[u], Sin[u], 0}, {u, 0,
2*Pi}, Axes -> False, Boxed -> False,
PlotStyle  -> Red, PlotRange -> {{-2, 2},
{-2, 2}, {0, 4}}]]}]
```

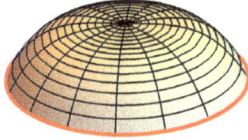

Out[ ]=

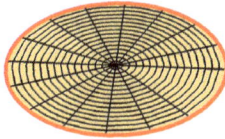

Figure 6.4

Note that a surface can also be piecewise smooth, i. e., it can be divided into a finite number of smooth pieces (the overlapping parts of the boundary of adjacent pieces will have opposite orientation).

6.4.1 Example: orientation of the boundary for Stokes' theorem

Let us study the orientation of the upper hemisphere. The unit upper hemisphere can be parametrized by

$$x = \sin u \cos v, \ y = \sin u \sin v, \ z = \cos u,$$

where $0 \le u \le \pi/2$ and $0 \le v \le 2\pi$. A normal vector to it is given by the cross-product of two tangent vectors

```
In[·]:= norm[{x_, y_, z_}] := Simplify[Sqrt[x^2 +
        y^2 + z^2]];
        N1[u_, v_] := Simplify[(Cross[D[#1, u],
        D[#1, v]] & )[{Sin[u]*Cos[v], Sin[u]*Sin[v],
        Cos[u]}]]
```

and a unit normal vector is

```
In[·]:= n1[u_, v_] := Assuming[u >= 0 && u <=
        Pi/2, Refine[Simplify[N1[u, v]/norm[N1[u, v]]]]]
```

```
In[·]:= n1[u, v]
Out[·]:= {Cos[v] Sin[u], Sin[u] Sin[v], Cos[u]}
```

Suppose that the unit hemisphere is oriented by the outward pointing normal. The correct orientation of the boundary circle can be visualized in the following way: the blue point (which is drawn a bit away from the boundary) moves along the boundary in such a direction that the normal vector is on the left (we visualize this by a small arrow from the blue point which is pointing towards the surface). In our example the correct orientation is counterclockwise.

```
In[·]:= Manipulate[Show[{ParametricPlot3D[
        {Sin[u]*Cos[v], Sin[u]*Sin[v], Cos[u]},
        {u, 0, Pi/2}, {v, 0, 2*Pi}, Axes -> False,
        Boxed -> False, PlotStyle -> LightYellow,
        PlotRange -> {{-2, 2}, {-2, 2}, {-2, 2}}],
        ParametricPlot3D[{Sin[u]*Cos[v],
        Sin[u]*Sin[v], Cos[u]} /. {u -> Pi/2},
        {v, 0, 2*Pi}, Axes -> False, Boxed ->
        False, PlotStyle -> {Thick, Red}], Graphics3D[
        {PointSize[0.06], Blue, Point[{1.3*Sin[u]*
        Cos[t], 1.3*Sin[u]*Sin[t], (1/3)*Cos[u]} /.
        {u -> Pi/2}]}], Graphics3D[Arrow[{{1.3*
        Sin[u]*Cos[t], 1.3*Sin[u]*Sin[t], (1/3)*Cos[u]}
        /. {u -> Pi/2}, {Sin[u]*Cos[t], Sin[u]*
        Sin[t], Cos[u]} /. {u -> Pi/2}}]], Graphics3D[
        Arrow[{{Sin[u]*Cos[v], Sin[u]*Sin[v], Cos[u]},
        n1[u, v] + {Sin[u]*Cos[v], Sin[u]*Sin[v],
        Cos[u]}} /. {u -> Pi/4, v -> 3*(Pi/4)}]]}],
        {t, 0, 2*Pi}, SaveDefinitions -> True]
```

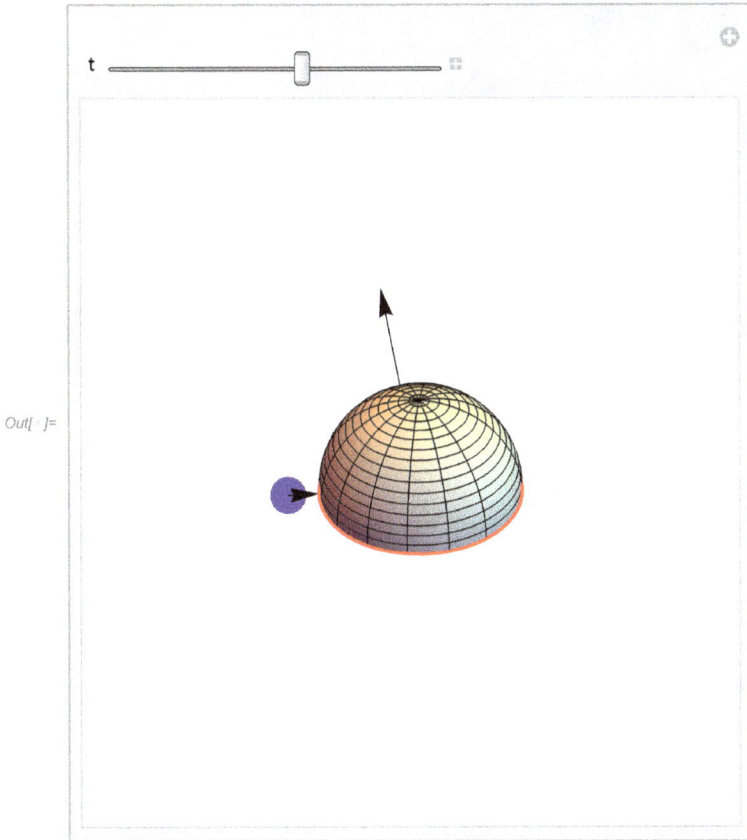

Figure 6.5

In the following example the direction is counterclockwise according to the orientation rule if the blue point is inside of the bowl and the normal vector points up.

```
In[·]:= N2[u_, v_] := Simplify[(Cross[D[#1, u],
        D[#1, v]] & )[{Sin[u]*Cos[v], Sin[u]*
        Sin[v], -Cos[u]}]];
        n2[u_, v_] := Assuming[u >= Pi/2 &&
        u <= Pi, Refine[Simplify[N2[u, v]/
        norm[N2[u, v]]]]]; Show[{ParametricPlot3D[
        {Sin[u]*Cos[v], Sin[u]*Sin[v], -Cos[u]},
        {u, 0, Pi/2}, {v, 0, 2*Pi},  Axes -> False,
        Boxed -> False, PlotStyle -> LightYellow,
        PlotRange -> {{-2, 2}, {-2, 2}, {-2, 2}}],
        ParametricPlot3D[{Sin[u]*Cos[v],
        Sin[u]*Sin[v], -Cos[u]} /. {u -> Pi/2},
        {v, 0, 2*Pi}, Axes -> False, Boxed -> False,
```

```
PlotStyle -> {Thick, Red}], Graphics3D[
 {PointSize[0.06], Blue, Point[{0.85*
Sin[u]*Cos[0], 0.85*Sin[u]*Sin[0],
-0.85*Cos[u]} /. {u -> Pi/2}]}],
Graphics3D[Arrow[{{Sin[u]*Cos[v],
Sin[u]*Sin[v], -Cos[u]}, n2[u, v] +
{Sin[u]*Cos[v], Sin[u]*Sin[v],
-Cos[u]}} /. {u -> Pi/4, v -> 3*(Pi/4)}]]}]
```

Out[ ]=

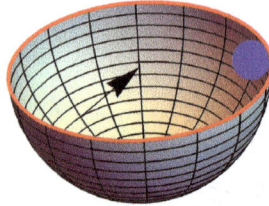

Figure 6.6

In the following example the orientation is clockwise according to the orientation rule if the blue point is outside of the lower bowl with the normal pointing downwards.

```
In[·]:= Show[{ParametricPlot3D[{Sin[u]*Cos[v],
Sin[u]*Sin[v], -Cos[u]}, {u, 0, Pi/2},
{v, 0, 2*Pi},  Axes -> False, Boxed ->
False, PlotStyle -> LightYellow,
PlotRange -> {{-2, 2}, {-2, 2},
{-2, 2}}], ParametricPlot3D[
{Sin[u]*Cos[v], Sin[u]*Sin[v], -Cos[u]}
/. {u -> Pi/2}, {v, 0, 2*Pi}, Axes -> False,
Boxed -> False, PlotStyle -> {Thick, Red}],
Graphics3D[{PointSize[0.06], Blue,
Point[{1.25*Sin[u]*Cos[0], 1.25*Sin[u]*
Sin[0], -1.25*Cos[u]} /. {u -> Pi/2}]}],
Graphics3D[Arrow[{{Sin[u]*Cos[v],
Sin[u]*Sin[v], -Cos[u]}, -n2[u, v] +
{Sin[u]*Cos[v], Sin[u]*Sin[v], -Cos[u]}}
/. {u -> Pi/4, v -> Pi/4}]]}]
```

Out[ ]=

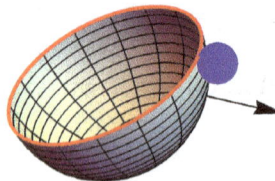

Figure 6.7

6.5 Stokes' theorem for graphs and its proof

Theorem 14 (Stokes' theorem). *Let S be an oriented surface parametrized by $(x, y, z = f(x, y))$, $(x, y) \in D$, where f is a C^2 function and D is a region to which Green's theorem can be applied, and let **F** be a C^1 vector field on S. Then*

$$\iint_S \text{curl } \mathbf{F} \cdot d\mathbf{S} = \iint_S (\nabla \times \mathbf{F}) \cdot d\mathbf{S} = \int_{\partial S} \mathbf{F} \cdot d\mathbf{s}.$$

The idea of the proof is to express integrals on both sides in terms of double integrals over D by using Green's theorem.

Let the vector field **F** be given by

In[·]:= F[{x_, y_, z_}] := {F1[x, y, z],
        F2[x, y, z], F3[x, y, z]}

The curl is

In[·]:= Curl[F[{x, y, z}], {x, y, z}]
Out[·]:= {− F2$^{(0,0,1)}$[x, y, z] + F3$^{(0,1,0)}$[x, y, z],
        F1$^{(0,0,1)}$[x, y, z] − F3$^{(1,0,0)}$[x, y, z],
        − F1$^{(0,1,0)}$[x, y, z] + F2$^{(1,0,0)}$[x, y, z]}

Here $F1^{(0,0,1)}(x, y, z)$ stands for the partial derivative with respect to z.

Let us recall some of the assumptions of the Stokes theorem. The surface S is given by $z = f(x, y)$, where (x, y) are in D. The function f is of class C^2. The boundary of D, that is, ∂D, is parametrized by $(x(t), y(t))$. Assume that Green's theorem applies to this region. The boundary of S, the closed curve C, is then parametrized by $(x(t), y(t), f(x(t), y(t)))$. Assume that the orientations of the boundaries and of the surface are all compatible.

To calculate $\iint_S \text{curl } \mathbf{F} \cdot d\mathbf{S}$, we need to express it as a double integral $\iint_D (\text{expr}_1)\, dA$, where $\text{expr}_1 = \text{curl } \mathbf{F}(x, y, f(x, y)) \cdot (-\frac{\partial f}{\partial x}, -\frac{\partial f}{\partial y}, 1)$:

In[·]:= expr1 = Curl[F[{x, y, f[x, y]}], {x, y, z}].
        ({-D[f[x, y], x], -D[f[x, y], y], 1}) // Simplify;

We omit the long outputs. To calculate the line integral $\int_{\partial S} \mathbf{F} \cdot d\mathbf{s}$, we substitute the parametrization of ∂S into **F** and then compute $\text{expr}_2(t)$ – its dot product with $(x'(t), y'(t), \frac{d}{dt} f(x(t), y(t)))$. The line integral is then $\int_a^b \text{expr}_2(t) dt$, where $t \in [a, b]$.

In[·]:= F[{x[t], y[t], f[x[t], y[t]]}] .
        {Derivative[1][x][t], Derivative[1][y][t],
        D[f[x[t], y[t]], t]};

After rearranging the terms, we have

In[·]:= expr2 = Collect[%, {Derivative[1][x][t],
        Derivative[1][y][t]}];

The integral with respect to t of this expression can be recognized as the line integral of the new vector field $(P(x, y), Q(x, y))$ over the boundary of the region D:

$$\int_{\partial D} (P, Q) \cdot d\mathbf{s} = \int_a^b (P(x(t), y(t)) x'(t) + Q(x(t), y(t)) y'(t)) \, dt$$

for the functions

In[·]:= P[x_, y_] := F1[x, y, f[x, y]] +
         F3[x, y, f[x, y]] D[f[x, y], x]

In[·]:= Q[x_, y_] := F2[x, y, f[x, y]] +
         F3[x, y, f[x, y]] D[f[x, y], y]

and, thus, Green's theorem gives the double integral $\iint_D (\mathrm{expr}_3) \, dA$, where $\mathrm{expr}_3 = \frac{\partial Q}{\partial x} - \frac{\partial P}{\partial y}$ (here we need the function f to be of class C^2):

In[·]:= expr3 = D[Q[x, y], x] - D[P[x, y], y] // Simplify;

Now we have two double integrals $\iint_D (\mathrm{expr}_1) \, dA$ and $\iint_D (\mathrm{expr}_3) \, dA$ (coming from the surface integral and from the line integral) and the expressions under the integrals are equal:

In[·]:= expr1 - expr3
Out[·]:= 0

This proves the statement of the theorem.

6.6 Stokes' theorem for parametrized surfaces and its proof

Theorem 15 (Stokes' theorem). *Let S be an oriented surface defined by a one-to-one parametrization $\Phi : D \subset \mathbb{R}^2 \to S$, where D is a region to which Green's theorem applies. Let ∂S denote the oriented boundary of S and let $\mathbf{F}$ be a C^1 vector field on S. Then*

$$\int_{\partial S} \mathbf{F} \cdot d\mathbf{s} = \iint_S \mathrm{curl}\, \mathbf{F} \cdot d\mathbf{S}.$$

If S has no boundary, then $\iint_S \mathrm{curl}\, \mathbf{F} \cdot d\mathbf{S} = 0$.

The idea of the proof is as follows. Suppose that the surface S is given by $\mathbf{r}(u, v) = (r_1(u, v), r_2(u, v), r_3(u, v))$ (one-to-one map), where (u, v) is in D. The boundary of D, that is, ∂D, is parametrized by $(u(t), v(t))$. Assume that Green's theorem applies to this region. The boundary of S, the closed curve C, is parametrized by $(r_1(u(t), v(t)), r_2(u(t), v(t)), r_3(u(t), v(t)))$. Assume that the orientations of the boundaries and of the surface are all compatible.

To calculate $\iint_S \text{curl } \mathbf{F} \cdot d\mathbf{S}$, we need to express it as a double integral $\iint_D (\text{expr}_4) \, dA$, where $\text{expr}_4 = \text{curl } \mathbf{F}(r_1(u,v), r_2(u,v), r_3(u,v)) \cdot (T_u \times T_v)$, where

$$T_u = \frac{\partial}{\partial u}(r_1(u,v), r_2(u,v), r_3(u,v)), \quad T_v = \frac{\partial}{\partial v}(r_1(u,v), r_2(u,v), r_3(u,v)).$$

In[·]:= Curl[F[{x, y, z}], {x, y, z}] /. {x -> r1[u, v],
      y -> r2[u, v], z -> r3[u, v]};

In[·]:= expr4 = Simplify[% . Cross[D[{r1[u, v], r2[u, v],
      r3[u, v]}, u], D[{r1[u, v], r2[u, v], r3[u, v]},
      v]]];

To calculate the line integral $\int_{\partial S} \mathbf{F} \cdot d\mathbf{s}$, we substitute the parametrization of ∂S into $\mathbf{F}$ and, as before, compute its dot product with $\frac{d}{dt}(r_1(u(t), v(t)), r_2(u(t), v(t)), r_3(u(t), v(t)))$ obtaining $\text{expr}_5(t)$. The line integral is then $\int_a^b \text{expr}_5(t) \, dt$, where $t \in [a, b]$. We omit the long outputs.

In[·]:= F[{r1[u[t], v[t]], r2[u[t], v[t]], r3[u[t], v[t]]}] .
      D[{r1[u[t], v[t]], r2[u[t], v[t]], r3[u[t],
      v[t]]}, t];

In[·]:= expr5 = Collect[%, {Derivative[1][u][t],
      Derivative[1][v][t]}];

The integral with respect to t of this expression can be recognized as the line integral of a new vector field $(P(u,v), Q(u,v))$ over the boundary of the region D:

$$\int_{\partial D} (P, Q) \cdot d\mathbf{s} = \int (P(u(t), v(t))u'(t) + Q(u(t), v(t))v'(t)) \, dt$$

for the functions

In[·]:= P[u_, v_] := F1[r1[u, v], r2[u, v], r3[u, v]]*
      D[r1[u, v], u] + F2[r1[u, v], r2[u, v], r3[u, v]]*
      D[r2[u, v], u] + F3[r1[u, v], r2[u, v], r3[u, v]]*
      D[r3[u, v], u]

In[·]:= Q[u_, v_] := F1[r1[u, v], r2[u, v], r3[u, v]]*
      D[r1[u, v], v] + F2[r1[u, v], r2[u, v], r3[u, v]]*
      D[r2[u, v], v] + F3[r1[u, v], r2[u, v], r3[u, v]]*
      D[r3[u, v], v]

and, so, Green's theorem gives the double integral $\iint_D (\text{expr}_6) \, dA$, where $\text{expr}_6 = Q_u - P_v$:

In[·]:= expr6 = Simplify[D[Q[u, v], u] - D[P[u, v], v]];

Now we have two double integrals $\iint_D (\text{expr}_4) \, dA$ and $\iint_D (\text{expr}_6) \, dA$ (coming from the line integral and from the surface integral) and the expressions under the integrals are equal:

In[·]:= expr4 - expr6 // Simplify
Out[·]:= 0

This proves the statement of the theorem.

6.7 Example 1

In this example we show that we need to check assumptions in Stokes' theorem carefully. Let S be a unit disk in the xy-plane and the vector field be given by the formula

In[·]:= F[x_, y_, z_] := {-y/(x^2 + y^2),
        x/(x^2 + y^2), 0}

The surface integral on the right-hand side of Stokes' theorem is zero, since the curl is zero:

In[·]:= Curl[F[x, y, z], {x, y, z}] // Simplify
Out[·]:= {0, 0, 0}

However, the line integral over the boundary of S (which is the unit circle on the xy-plane) on the left-hand side of Stokes' formula is not zero:

In[·]:= Integrate[Simplify[F[Cos[t], Sin[t], 0] .
        D[{Cos[t], Sin[t], 0}, t]], {y, 0, 2*Pi}]
Out[·]:= 2π

Note that in this case the vector field is not defined on the whole S (in fact, it is not defined on the whole z-axis, i. e., at points $(0, 0, z)$).

6.8 Example 2

Let us show that the vector field

In[·]:= V[x1_, x2_, x3_] := {-2*x1*(x3/(x3^2
        + (x1^2 + x2^2 - 1)^2)), -2*x2*(x3/(x3^2 +
        (x1^2 + x2^2 - 1)^2)), (x1^2 + x2^2 - 1)/
        (x3^2 + (x1^2 + x2^2 - 1)^2)}

is curl-free, but the integration along the path $(\sqrt{1 + \cos t}, 0, \sin t)$, $-\pi \le t \le \pi$, is not zero.

In[·]:= Curl[V[x1, x2, x3], {x1, x2, x3}] // Simplify
Out[·]:= {0, 0, 0}

In[·]:= Integrate[Simplify[V[Sqrt[1 + Cos[t]],
        0, Sin[t]] . D[{Sqrt[1 + Cos[t]], 0, Sin[t]},
        t]], {t, -Pi, Pi}]
Out[·]:= 2π

```
In[·]:= ParametricPlot3D[{Sqrt[1 + Cos[t]], 0, Sin[t]},
         {t, -Pi, Pi}]
```

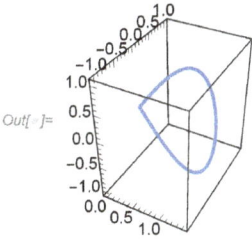

Figure 6.8

6.9 Example 3

In this example we show that the computation of the flux integral in Stokes' theorem is sometimes easier than the computation of the line integral. Let us check Stokes' theorem for the vector field $\mathbf{F} = (xy, xy + x, z)$ with curl $\mathbf{F} = \mathbf{H} = (0, 0, 1 - x + y)$. The closed curve $x(t) = \cos t + 2$, $y(t) = \sin t + 2$, $z(t) = \sin((\cos t + 2)^2 + (\sin t + 2)^2) + 2$, $0 \leq t \leq 2\pi$, oriented counterclockwise, lies on the surface $z = \sin(x^2 + y^2) + 2$ oriented by the normal pointed upwards. The curve and part of the surface is visualized as

```
In[·]:= Show[ParametricPlot3D[{r*Cos[t] + 2,
         r*Sin[t] + 2, Sin[(2 + r*Cos[t])^2 +
         (2 + r*Sin[t])^2] + 2}, {r, 0, 1}, {t, 0, 2*Pi},
         PlotStyle -> LightYellow], ParametricPlot3D[{Cos[t] +
         2, Sin[t] + 2, Sin[(Cos[t] + 2)^2 + (Sin[t] + 2)^2]
         + 2}, {t, 0, 2*Pi}, PlotStyle -> Red]]
```

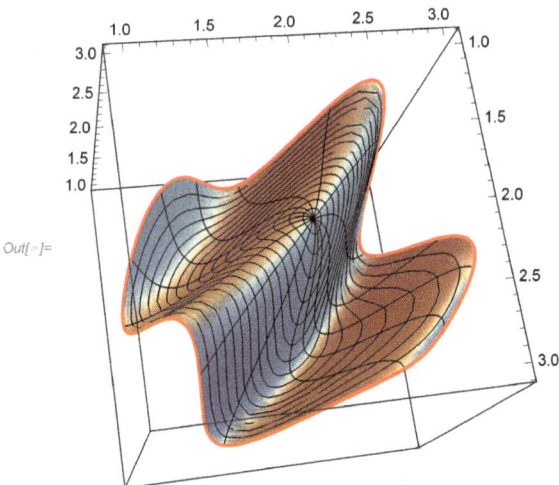

Figure 6.9

In[·]:= F[{x_, y_, z_}] := {x y, x y + x, z}

In[·]:= Curl[F[{x, y, z}], {x, y, z}]
Out[·]:= {0, 0, 1 - x + y}

In[·]:= H[{x_, y_, z_}] := {0, 0, 1 - x + y}

The line integral in Stokes' theorem can be evaluated as follows:

In[·]:= Integrate[Simplify[F[{Cos[t] + 2, Sin[t] + 2,
          Sin[(Cos[t] + 2)^2 + (Sin[t] + 2)^2] + 2}] .
          D[{Cos[t] + 2, Sin[t] + 2, Sin[(Cos[t] + 2)^2 +
          (Sin[t] + 2)^2] + 2}, t]], {t, 0, 2*Pi}]
Out[·]:= π

In this example the calculation of the line integral is difficult to do by hand, whereas the flux integral will be much easier. The flux integral in Stokes' theorem can be evaluated as follows (after changing to generalized polar coordinates and taking into account the Jacobian of the change of variables in the calculation of the double integral):

In[·]:= Simplify[H[{x, y, Sin[x^2 + y^2] + 2}] .
          {-D[Sin[x^2 + y^2] + 2, x], -D[Sin[x^2 + y^2] + 2,
          y], 1}]
Out[·]:= 1 - x + y

In[·]:= Simplify[% /. {x -> r*Cos[t] + 2,
          y -> r*Sin[t] + 2}]
Out[·]:= 1 - r Cos[t] + r Sin[t]

In[·]:= Integrate[%*Det[{{D[r*Cos[t] + 2, r],
          D[r*Cos[t] + 2, t]}, {D[r*Sin[t] + 2, r],
          D[r*Sin[t] + 2, t]}}], {r, 0, 1}, {t, 0, 2*Pi}]
Out[·]:= π

6.10 Example 4

In this example we verify Stokes' theorem for a unit circle in the xy-plane and for two surfaces that share the same boundary (to show surface independence of Stokes' theorem). We shall check Stokes' theorem for the curve C which is a unit circle in xy-plane, orientated counterclockwise. The first surface – the upper part of the ellipsoid – is parametrized by $x(u,v) = \sin u \cos v$, $y(u,v) = \sin u \sin v$, $z(u,v) = (\cos u)/2$, where $0 \le u \le \pi/2$, $0 \le v \le 2\pi$. The second surface is a unit disk in the xy-plane. We choose outwards (upwards) pointing normal for the orientation of the surfaces. The vector field is $\mathbf{F} = (x-y+x^2y^2, xy+y^2+z^2, x^2+y^2+z^3)$ with curl $\mathbf{F} = \mathbf{G} = (2y-2z, -2x, 1+y-2x^2y)$.

In[·]:= F[{x_, y_, z_}] := {x - y + x^2*y^2,
    x*y + y^2 + z^2, x^2 + y^2 + z^3};

In[·]:= Curl[F[{x, y, z}], {x, y, z}]
Out[·]:= {2 y - 2 z, -2 x, 1 + y - 2 x^2 y

The first surface with an outward pointing normal is as follows:

In[·]:= circle = ParametricPlot3D[{Cos[u], Sin[v],
    0}, {v, 0, 2*Pi}, Boxed -> False, Axes -> False,
    PlotRange -> {{-1, 1}, {-1, 1}, {0, 2}},
    PlotStyle -> Blue];
    surf1 = Show[{Graphics3D[Opacity[0.3],
    Boxed -> False], ParametricPlot3D[
    {Sin[u]*Cos[v], Sin[u]*Sin[v], Cos[u]/2},
    {v, 0, 2*Pi}, {u, 0, Pi/2}, Axes -> False,
    PlotStyle -> Pink, PlotRange -> {{-1, 1},
    {-1, 1}, {0, 1}}]}]; N11[u_, v_] :=
    Simplify[(Cross[D[#1, u], D[#1, v]] &)
    [{Sin[u]*Cos[v], Sin[u]*Sin[v], Cos[u]/2}]];
    n11[u_, v_] := Assuming[u >= 0 && u <= Pi/2,
    Refine[Simplify[N1[u, v]/norm[N1[u, v]]]]];
    surf1norm = Show[{Graphics3D[Opacity[0.3],
    Boxed -> False],
    ParametricPlot3D[{Sin[u]*Cos[v],
    Sin[u]*Sin[v], Cos[u]/2}, {v, 0, 2*Pi},
    {u, 0, Pi/2}, Boxed -> False, Axes -> False,
    PlotStyle -> Pink, PlotRange -> {{-1, 1},
    {-1, 1}, {0, 1}}], Graphics3D[Opacity[1]],
    Graphics3D[Arrow[{{Sin[u]*Cos[v],
    Sin[u]*Sin[v], Cos[u]/2}, n11[u, v] +
    {Sin[u]*Cos[v], Sin[u]*Sin[v], Cos[u]/2}}
    /. {u -> Pi/4, v -> 3*(Pi/4)}]]}]

Out[·]=

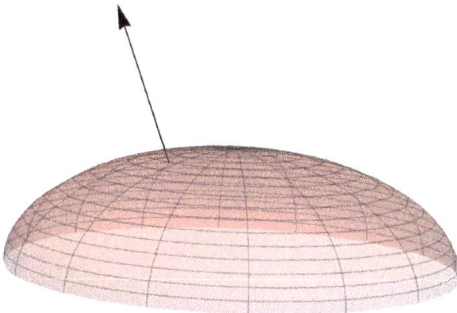

Figure 6.10

The second surface with the upward pointing normal can be visualized as follows (note the order of the cross-product of two tangent vectors for the pointing upwards normal):

```
In[·]:= N2[u_, v_] := Simplify[(Cross[D[#1, v],
        D[#1, u]] & )[{v*Cos[u], v*Sin[u], 0}]];
        n2[u_, v_] := Assuming[u >= 0 && u <= 2*Pi
        && v >= 0, Refine[Simplify[N2[u, v]/
        norm[N2[u, v]]]]]; surf2 = RegionPlot3D[
        x^2 + y^2 <= 1 && z == 0, {x, -1, 1},
        {y, -1, 1}, {z, 0, 1}, Axes -> False,
        Boxed -> False]; surf2norm = Show[
        {Graphics3D[Opacity[0.3], Boxed -> False],
        ParametricPlot3D[{v*Cos[u], v*Sin[u],
        0}, {u, 0, 2*Pi}, {v, 0, 1}, Boxed ->
        False, Axes -> False, PlotStyle ->
        Yellow, PlotRange -> {{-1, 1}, {-1, 1},
        {0, 1}}], Graphics3D[Opacity[1]],
        Graphics3D[Arrow[{{v*Cos[u], v*Sin[u], 0},
        n2[u, v] + {v*Cos[u], v*Sin[u], 0}}
        /. {u -> 0, v -> 0}]]}]
```

Out[·]=

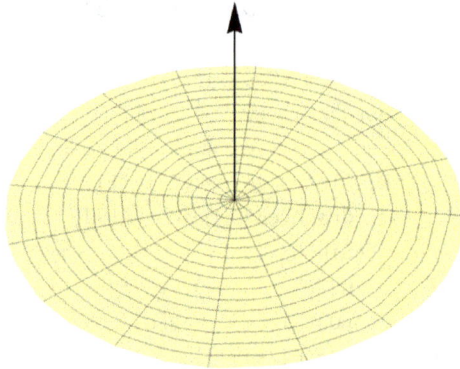

Figure 6.11

The surfaces share the same boundary:

```
In[·]:= Show[surf1, surf2, circle]
```

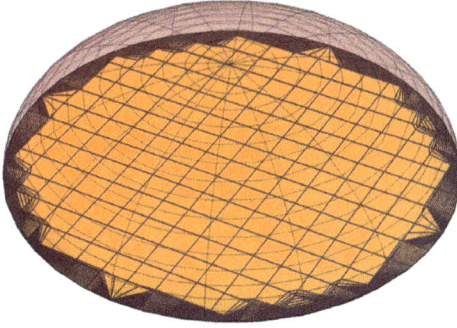

Out[ ]=

Figure 6.12

To calculate the line integral in Stokes' theorem, we use the parametrization of the circle $x(u) = \cos u$, $y(u) = \sin u$, $z(u) = 0$, $0 \leq u \leq 2\pi$:

```
In[·]:= Integrate[(F[{x, y, z}] /. {x -> Cos[u],
          y -> Sin[u], z -> 0}) . D[{Cos[u], Sin[u], 0},
          u], {u, 0, 2*Pi}]
```
Out[·]:= π

Calculating the flux integrals for two surfaces in Stokes' theorem, we get

```
In[·]:= surfpar1[u_, v_] := {Sin[u]*Cos[v],
          Sin[u]*Sin[v], Cos[u]/2};
          surfpar2[u_, v_] := {v*Cos[u], v*Sin[u], 0};
          G[{x_, y_, z_}] := Simplify[{2*y - 2*z,
          -2*x, 1 + y - 2*x^2*y}];

In[·]:= Integrate[Simplify[G[surfpar1[u, v]] .
          (Cross[D[#1, u], D[#1, v]] & )[surfpar1[u, v]]],
          {v, 0, 2*Pi}, {u, 0, Pi/2}]
```
Out[·]:= π

```
In[·]:= Integrate[Simplify[G[surfpar2[u, v]] .
          (Cross[D[#1, v], D[#1, u]] & )[surfpar2[u, v]]],
          {u, 0, 2*Pi}, {v, 0, 1}]
```
Out[·]:= π

Bibliography

[1] Bochnak J, Coste M, Roy M-F. Real Algebraic Geometry. A Series of Modern Surveys in Mathematics 36, Springer-Verlag Berlin Heidelberg GmbH, 1998.

[2] Canuto C, Tabacco A. Mathematical Analysis II. Springer, 2010.

[3] Cox D, Little J, O'Shea D. Ideals, Varieties and Algorithms. An Introduction to Computational Algebraic Geometry and Commutative Algebra. Undergraduate Texts in Mathematics, Third edition. Springer Science+Business Media, LLC, 2007.

[4] Edelsbrunner H, Harer JL. Computational Topology, An Introduction. American Mathematical Society, Providence RI, 2010.

[5] Filipuk G, Kozłowksi A. Analysis with Mathematica®. Volume 1: Single Variable Calculus. De Gruyter Textbook, de Gruyter, 2019.

[6] Katz V. The history of Stokes' theorem. Mathematics Magazine 52 (3) (1979), 146–156.

[7] Körner TW. A Companion to Analysis. A Second First and First Second Course in Analysis. Graduate Studies in Mathematics 62, American Mathematical Society, Providence, RI, 2004.

[8] Kozłowski A. Morse-Smale flows on a tilted torus. Wolfram Demonstrations Project. Published: March 7 2011. Available at http://demonstrations.wolfram.com/MorseSmaleFlowsOnATiltedTorus/.

[9] Madsen I, Tornehave J. From Calculus to Cohomology. Cambridge University Press, 1997.

[10] Marsden J, Tromba A. Vector Calculus, 6th edition. W.H. Freeman and Company, New York, 2012.

[11] Milnor J. Morse Theory. Annals of Mathematics Studies 51, Princeton University Press, 1973.

[12] Moszyński M. Analiza matematyczna dla informatyków. Wykłady dla pierwszego roku informatyki na Wydziale Matematyki, Informatyki i Mechaniki Uniwersytetu Warszawskiego, 2010 (in Polish). Available at https://www.mimuw.edu.pl/~mmoszyns/Analiza-dla-Informatykow-2017-18/SKRYPT.

[13] Spivak M. Calculus on Manifolds. A Modern Approach to Classical Theorems of Advanced Calculus. Addison-Wesley Publishing Company, The Advanced Book Program, US, 1965.

[14] Tao T. Analysis II, Third edition. Texts and Readings in Mathematics 38, Springer, Hindustan Book Agency, 2015.

[15] Thomson BS, Bruckner JB, Bruckner AM. Elementary Real Analysis, Second edition, CreateSpace Independent Publishing Platform, 2008. Available at ClassicalRealAnalysis.com.

[16] Vanyo J. Rotating Fluids in Engineering and Science. Butterworth – Heinemann, 2015.

https://doi.org/10.1515/9783110660395-007

Index